ST(P) MATHEMATICS 2

ST(P) MATHEMATICS will be completed as follows:

Published 1984
- **ST(P) 1**
- **ST(P) 1**   Teacher's Notes and Answers
- **ST(P) 2**

Published 1985
- **ST(P) 2**   Teacher's Notes and Answers
- **ST(P) 3A**
- **ST(P) 3B**
- **ST(P) 3A**   Teacher's Notes and Answers
- **ST(P) 3B**   Teacher's Notes and Answers

Published 1986
- **ST(P) 4A**
- **ST(P) 4B**
- **ST(P) 4A**   Teacher's Notes and Answers
- **ST(P) 4B**   Teacher's Notes and Answers

Published 1987
- **ST(P) 5A**   (with answers)
- **ST(P) 5B**   (with answers)

Published 1988
- **ST(P) 5C**   (with answers)

In preparation
- **ST(P)** Resource Book

ST(P) MATHEMATICS 2

L. Bostock, B.Sc.

formerly Senior Mathematics Lecturer, Southgate Technical College

S. Chandler, B.Sc.

formerly of the Godolphin and Latymer School

A. Shepherd, B.Sc.

Head of Mathematics, Redland High School for Girls

E. Smith, M.Sc.

Head of Mathematics, Tredegar Comprehensive School

Stanley Thornes (Publishers) Ltd

First published 1984 by
Stanley Thornes (Publishers) Ltd,
Old Station Drive,
Leckhampton,
CHELTENHAM GL53 0DN

Reprinted 1985
Reprinted 1986 (twice)
Reprinted 1987
Reprinted 1988

British Library Cataloguing in Publication Data
ST(P) mathematics
    Book 2
    1. Mathematics——1961—
    I. Bostock, L.
    510   QA39.2

ISBN 0–85950–174–4

Typeset by Cotswold Typesetting Ltd, Gloucester
Printed and bound in Great Britain at The Bath Press, Avon

CONTENTS

INTRODUCTION

This book continues the attempt to satisfy your needs as you study mathematics in the secondary school. We are conscious of the need for success together with the enjoyment everyone finds in getting things right. With this in mind we have provided plenty of straightforward questions and have divided the exercises into three types of question:

The first type, identified by plain numbers, e.g. **12.**, helps you to see if you understand the work. These questions are considered necessary for every chapter you attempt.

The second type, identified by a single underline, e.g. **12.**, are extra, but not harder, questions for quicker workers, for extra practice or for later revision.

The third type, identified by a double underline, e.g. **12.**, are for those of you who manage Type 1 questions fairly easily and therefore need to attempt questions that are a little harder.

Most chapters end with "mixed exercises". These will help you revise what you have done, either when you have finished the chapter or at a later date.

At this stage you will find that you use your calculator more frequently. However, it is still wise to use it mainly to check answers. Whether you use a calculator or not, always estimate your answer and always ask yourself the question, "Is my answer a sensible one?"

1 WORKING WITH NUMBERS

POSITIVE INDICES

We have seen that 3^2 means 3×3
  and that $2 \times 2 \times 2$ can be written as 2^3.
The small number at the top is called the *index* or *power*. (The plural of index is indices.)
It follows that 2 can be written as 2^1 although we would not normally do so.

$$5^1 \quad \text{means} \quad 5$$

EXERCISE 1a

Find 2^5

$$2^5 = 2 \times 2 \times 2 \times 2 \times 2$$
$$= 32$$

Find:

1. 3^2	**4.** 5^3	**7.** 2^7	**10.** 10^4
2. 4^1	**5.** 10^3	**8.** 10^1	**11.** 10^6
3. 10^2	**6.** 3^4	**9.** 4^3	**12.** 3^3

Find the value of 3.6×10^2

$$3.6 \times 10^2 = 3.6 \times 100$$
$$= 360$$

Find the value of:

13. 7.2×10^3	**18.** 5.37×10^5
14. 8.93×10^2	**19.** 4.63×10^1
15. 6.5×10^4	**20.** 5.032×10^2
16. 3.82×10^3	**21.** 7.09×10^2
17. 2.75×10^1	**22.** 6.978×10^1

1

MULTIPLYING NUMBERS WRITTEN IN INDEX FORM

We can write $2^2 \times 2^3$ as a single number in index form because

$$2^2 \times 2^3 = (2 \times 2) \times (2 \times 2 \times 2)$$
$$= 2 \times 2 \times 2 \times 2 \times 2$$
$$= 2^5$$
$$\therefore \quad 2^2 \times 2^3 = 2^{2+3} = 2^5$$

But we cannot do the same with $2^2 \times 5^3$ because the numbers multiplied together are not all 2s (nor are they all 5s).

We can multiply together different powers of the *same* number by adding the indices but we cannot multiply together powers of different numbers in this way.

EXERCISE 1b

Write $a^3 \times a^4$ as a single expression in index form.

$$a^3 \times a^4 = a^{3+4}$$
$$= a^7$$

Write as a single expression in index form:

1. $3^5 \times 3^2$

2. $7^5 \times 7^3$

3. $9^2 \times 9^8$

4. $2^4 \times 2^7$

5. $b^3 \times b^2$

6. $5^4 \times 5^4$

7. $12^4 \times 12^5$

8. $p^6 \times p^8$

9. $4^7 \times 4^9$

10. $r^5 \times r^3$

DIVIDING NUMBERS WRITTEN IN INDEX FORM

If we want to write $2^5 \div 2^2$ as a single number in index form then

$$2^5 \div 2^2 = \frac{2^5}{2^2} = \frac{\cancel{2} \times \cancel{2} \times 2 \times 2 \times 2}{\cancel{2} \times \cancel{2}} = 2^3$$

i.e.
$$\frac{2^5}{2^2} = 2^{5-2} = 2^3$$

We can divide different powers of the *same* number by subtracting the indices.

EXERCISE 1c

> Write $a^7 \div a^3$ as a single expression in index form.
>
> $$a^7 \div a^3 = a^{7-3}$$
> $$= a^4$$

Write as a single expression in index form:

1. $4^4 \div 4^2$ **6.** $15^8 \div 15^4$

2. $7^9 \div 7^3$ **7.** $6^{12} \div 6^7$

3. $5^6 \div 5^5$ **8.** $b^7 \div b^5$

4. $10^8 \div 10^3$ **9.** $9^{15} \div 9^{14}$

5. $q^9 \div q^5$ **10.** $p^4 \div p^3$

11. $6^4 \times 6^7$ **16.** $2^2 \times 2^4 \times 2^3$

12. $3^9 \div 3^6$ **17.** $4^2 \times 4^3 \div 4^4$

13. $2^8 \div 2^7$ **18.** $a^2 \times a^2 \div a^3$

14. $a^9 \times a^3$ **19.** $3^6 \div 3^2 \times 3^4$

15. $c^6 \div c^3$ **20.** $b^2 \times b^3 \times b^4$

NEGATIVE INDICES

Consider $2^3 \div 2^5$

Subtracting the indices gives $2^3 \div 2^5 = 2^{3-5} = 2^{-2}$

But, as a fraction, $2^3 \div 2^5 = \dfrac{2^3}{2^5} = \dfrac{2 \times 2 \times 2}{2 \times 2 \times 2 \times 2 \times 2} = \dfrac{1}{2^2}$

Therefore 2^{-2} means $\dfrac{1}{2^2}$

In the same way, 5^{-3} means $\dfrac{1}{5^3}$

$\dfrac{1}{5^3}$ is called the *reciprocal* of 5^3, so also 5^{-3} is the reciprocal of 5^3

In general, a^{-b} is the *reciprocal* of a^b $\left(\text{i.e. } a^{-b} = \dfrac{1}{a^b} \right)$

EXERCISE 1d

Find the value of 5^{-2}

$$5^{-2} = \frac{1}{5^2}$$

$$= \frac{1}{25}$$

Find the value of:

1. 2^{-2}	**6.** 4^{-2}	**11.** 4^{-3}	**16.** 5^{-3}
2. 3^{-3}	**7.** 3^{-4}	**12.** 6^{-2}	**17.** 10^{-2}
3. 2^{-4}	**8.** 5^{-1}	**13.** 15^{-1}	**18.** 2^{-3}
4. 3^{-1}	**9.** 3^{-2}	**14.** 6^{-1}	**19.** 10^{-1}
5. 7^{-1}	**10.** 4^{-1}	**15.** 7^{-2}	**20.** 8^{-2}

Find the value of 1.7×10^{-2}

$$1.7 \times 10^{-2} = 1.7 \times \frac{1}{10^2}$$

$$= \frac{1.7}{100} = 0.017$$

Find the value of:

21. 3.4×10^{-3}	**26.** 4.67×10^{-5}
22. 2.6×10^{-1}	**27.** 3.063×10^{-1}
23. 6.2×10^{-2}	**28.** 2.805×10^{-2}
24. 8.21×10^{-3}	**29.** 51.73×10^{-4}
25. 5.38×10^{-4}	**30.** 30.04×10^{-1}

Write $2 \div 2^3$ as a single number in index form.

$$2 \div 2^3 = 2^1 \div 2^3$$

$$= 2^{-2}$$

Write as a single number in index form:

31. $5^2 \div 5^4$ **36.** $10^3 \div 10^6$

32. $3 \div 3^4$ **37.** $b^5 \div b^9$

33. $6^4 \div 6^7$ **38.** $4^8 \div 4^3$

34. $2^5 \div 2^3$ **39.** $c^5 \div c^4$

35. $a^5 \div a^7$ **40.** $2^a \div 2^b$

THE MEANING OF a^0

Consider $2^3 \div 2^3$

Subtracting indices gives $\qquad 2^3 \div 2^3 = 2^0$

Simplifying $\dfrac{2^3}{2^3}$ gives $\qquad \dfrac{\cancel{2} \times \cancel{2} \times \cancel{2}}{\cancel{2} \times \cancel{2} \times \cancel{2}} = 1$

So 2^0 means 1

In the same way $a^3 \div a^3 = a^0 \qquad$ (subtracting indices)

But $\qquad a^3 \div a^3 = \dfrac{a \times a \times a}{a \times a \times a} = 1 \qquad$ (simplifying the fraction)

> Any number with an index of zero is equal to 1
>
> i.e. $\quad a^0 = 1$

MIXED QUESTIONS ON INDICES

EXERCISE 1e   Find the value of:

1. 2^2 **4.** 3^{-1} **7.** 3^4 **10.** 6^{-2}

2. 5^{-2} **5.** 7^0 **8.** 2^0 **11.** 10^{-3}

3. 4^3 **6.** 5^3 **9.** 4^1 **12.** $\left(\dfrac{1}{2}\right)^{-1}$

13. 2.41×10^3 **18.** 1.074×10^{-1}

14. 7.032×10^{-1} **19.** 7.834×10^2

15. 4.971×10^2 **20.** 3.05×10^3

16. 7.805×10^{-3} **21.** 5.99×10^0

17. 5.92×10^4 **22.** 3.8601×10^{-4}

Write as a single number in index form:

23. $2^3 \times 2^4$

24. $4^6 \div 4^3$

25. $3^{-2} \times 3^4$

26. $a^4 \times a^3$

27. $a^7 \div a^3$

28. $5^4 \times 5^{-2}$

29. $3^5 \div 3^5$

30. $b^3 \div b^3$

31. $4^{-2} \times 4^6$

32. $5^3 \div 5^9$

33. $2^2 \times 2^4 \times 2^3$

34. $a^2 \times a^4 \times a^6$

35. $3^5 \times 3^2 \div 3^3$

36. $7^3 \times 7^3 \div 7^6$

37. $\dfrac{4^2 \times 4^6}{4^3}$

38. $a^3 \times a^2 \times a^5$

39. $3^2 \div 3^6 \times 3^2$

40. $b^3 \times b^{-3}$

41. $5^{-2} \times 5^{-3}$

42. $\dfrac{a^3 \times a^4}{a^7}$

STANDARD FORM

The nearest star to us (Alpha Centauri) is about 25 million million miles away. Written in figures this very large number is 25 000 000 000 000.

The diameter of an atom is roughly 2 ten-thousand-millionths of a metre, or 0.000 000 000 2 m and this is very small.

These numbers are cumbersome to write down and, until we have counted the zeros, we cannot tell their size. We need a way of writing such numbers in a shorter form from which it is easier to judge their size: the form that we use is called standard form (sometimes called scientific notation).

Written in standard form the first number is 2.5×10^{13}
and the second number is 2×10^{-10}

> Standard form is a number between 1 and 10
> multiplied by a power of 10.

So 1.3×10^2, 2.86×10^4 and 3.72×10^{-2} are in standard form,

but 13×10^3 and 0.36×10^{-2} are not in standard form because the first number is not between 1 and 10.

EXERCISE 1f

> Write 2.04×10^{-3} as an ordinary number.
>
> $$2.04 \times 10^{-3} = 2.04 \times \frac{1}{10^3}$$
>
> $$= 0.002\,04$$

Each of the following numbers is written in standard form. Write them as ordinary numbers.

1. 3.78×10^3 **6.** 3.67×10^{-6}

2. 1.26×10^{-3} **7.** 3.04×10^4

3. 5.3×10^6 **8.** 8.503×10^{-4}

4. 7.4×10^{14} **9.** 4.25×10^{12}

5. 1.3×10^{-4} **10.** 6.43×10^{-8}

CHANGING NUMBERS INTO STANDARD FORM

To change 6800 into standard form, the decimal point has to be placed between the 6 and the 8 to give a number between 1 and 10.
Counting then tells us that, to change 6.8 to 6800, we need to move the decimal point three places to the right (i.e. to multiply by 10^3)

i.e. $\qquad 6800 = 6.8 \times 1000 = 6.8 \times 10^3$

To change 0.019 34 into standard form, the point has to go between the 1 and the 9 to give a number between 1 and 10.
This time counting tells us that, to change 1.934 to 0.019 34, we need to move the point two places to the left (i.e. to divide by 10^2)

so $\qquad 0.019\,34 = 1.934 \div 100 = 1.934 \times 10^{-2}$

EXERCISE 1g Change the following numbers into standard form:

1. 2500   **6.** 39 070   **11.** 26 030

2. 630   **7.** 4 500 000   **12.** 547 000

3. 15 300   **8.** 530 000 000   **13.** 30 600

4. 260 000   **9.** 40 000   **14.** 4 060 000

5. 9900   **10.** 80 000 000 000   **15.** 704

Write 0.006 043 in standard form.

$$0.006\,043 = 6.043 \times 10^{-3}$$

Write the following numbers in standard form:

16.	0.026	**21.**	0.79	**26.**	0.907
17.	0.0048	**22.**	0.0069	**27.**	0.0805
18.	0.053	**23.**	0.000 007 5	**28.**	0.088 08
19.	0.000 018	**24.**	0.000 000 000 4	**29.**	0.000 704 4
20.	0.52	**25.**	0.684	**30.**	0.000 000 000 073

31.	79.3	**36.**	60.5	**41.**	5 300 000 000 000
32.	0.005 27	**37.**	0.003 005	**42.**	0.000 000 050 2
33.	80 600	**38.**	0.600 05	**43.**	0.007 008 09
34.	0.9906	**39.**	7 080 000	**44.**	708 000
35.	0.0705	**40.**	560 800	**45.**	40.5

46.	88.92	**51.**	84	**56.**	5090
47.	0.000 050 6	**52.**	351	**57.**	268 000
48.	0.000 000 057	**53.**	0.09	**58.**	30.7
49.	503 000 000	**54.**	0.007 05	**59.**	0.005 05
50.	99 000 000	**55.**	36	**60.**	0.000 008 8

APPROXIMATIONS: WHOLE NUMBERS

We saw in Book 1 that it is sometimes necessary to approximate given numbers by rounding them off to the nearest 10, 100, ... For example, if you measured your height in millimetres as 1678 mm, it would be reasonable to say that you were 1680 mm tall to the nearest 10 mm.

The rule is that if you are rounding off to the nearest 10 you look at the units. If there are 5 or more units you add one on to the tens. If there are less than 5 units you leave the tens alone.

Similar rules apply to rounding off to the nearest 100 (look at the tens); to the nearest 1000 (look at the hundreds); and so on.

EXERCISE 1h

Round off 1853 to   a) the nearest ten
                     b) the nearest hundred
                     c) the nearest thousand

a) $185\vert 3 = 1850$ to the nearest 10

b) $18\vert 53 = 1900$ to the nearest 100

c) $1\vert 853 = 2000$ to the nearest 1000

Round off each of the following numbers to   a) the nearest ten
b) the nearest hundred   c) the nearest thousand:

1.	1547	**5.**	68 414	**9.**	53 804
2.	8739	**6.**	5729	**10.**	6007
3.	2750	**7.**	4066	**11.**	4981
4.	36 835	**8.**	7507	**12.**	8699

A building firm stated that, to the nearest 100, it built 2600 homes last year. What is the greatest number of homes that it could have built and what is the least number of homes that it could have built?

The smallest number that can be rounded up to 2600 is 2550.
The biggest number that can be rounded down to 2600 is 2649.

So the firm built at most 2649 homes and at least 2550 homes.

13. A bag of marbles is said to contain 50 marbles to the nearest 10. What is the greatest number of marbles that could be in the bag and what is the least number of marbles that could be in the bag?

14. To the nearest thousand, the attendance at a particular First Division football match was 45 000. What is the largest number that could have been there and what is the smallest number that could have attended?

15. 1500 people came to the school fete. If this number is correct to the nearest hundred, give the maximum and the minimum number of people that could have come.

16. The annual accounts of Scrub plc (soap manufacturers) gave the company's profit as £3 000 000 to the nearest million. What is the least amount of profit that the company could have made?

17. The chairman of A. Brick (Builders) plc said that they employ 2000 people. If this number is correct to the nearest 100, what is the least number of employees that the company can have?

APPROXIMATIONS: DECIMALS

If you measure your height in centimetres as 167.8 cm, it would be reasonable to say that, to the nearest centimetre, you are 168 cm tall. We write   167.8 = 168   correct to the nearest unit.

If you measure your height in metres as 1.678 m, it would be reasonable to say that, to the nearest $\frac{1}{100}$ m, you are 1.68 m tall.
Hundredths are represented in the second decimal place so we say that   1.678 = 1.68   correct to 2 decimal places.

EXERCISE 1i

> Give 8.753 to a) 2 decimal places
> b) 1 decimal place
> c) the nearest unit
>
> a)   8.75$_|$3 = 8.75 correct to 2 d.p.
>
> b)   8.7$_|$53 = 8.8 correct to 1 d.p.
>
> c)   8$_|$.753 = 9   correct to the nearest unit

Give each of the following numbers correct to   a) 2 decimal places
b) 1 decimal place   c) the nearest unit:

1.	2.758	**6.**	3.896
2.	7.371	**7.**	8.936
3.	16.987	**8.**	73.649
4.	23.758	**9.**	6.896
5.	9.858	**10.**	55.575

Give the following numbers correct to the number of decimal places given in brackets:

11.	5.07	(1)		**16.**	0.9752	(3)
12.	0.0087	(3)		**17.**	5.5508	(3)
13.	7.897	(2)		**18.**	285.59	(1)
14.	34.82	(1)		**19.**	6.749	(1)
15.	0.007 831	(4)		**20.**	9.999	(2)

SIGNIFICANT FIGURES

In the previous two sections we used a height of 1678 mm as an example. This height was measured in three different units and then rounded off:

> in the first case to   1680 mm   correct to the nearest 10 mm,
> in the second case to 168 cm   correct to the nearest centimetre,
> in the third case to   1.68 m   correct to 2 d.p.

We could also give this measurement in kilometres, to the same degree of accuracy, as   0.001 68 km   correct to 5 d.p.

Notice that the three figures 1, 6 and 8 occur in all four numbers and that it is the 8 that has been corrected in each case.

The figures 1, 6 and 8 are called the *significant figures* and in all four cases the numbers are given correct to 3 significant figures.

Using significant figures rather than place values (i.e. tens, units, first d.p., second d.p., . . .) has advantages. For example, if you are asked to measure your height and give the answer correct to 3 significant figures, then you can choose any convenient unit. You do not need to be told which unit to use and which place value in that unit to correct your answer to.

Writing a number in standard form gives an easy way of finding the first significant figure: it is the number to the left of the decimal point.

For example $170.6 = \underline{1}.706 \times 10^2$

So 1 is the first significant figure in 170.6.

The second significant figure is the next figure to the right (7 in this case).

The third significant figure is the next figure to the right again (0 in this case), and so on.

EXERCISE 1j

> Write down   a) the first significant figure
> b) the third significant figure in 0.001 503
>
> $$0.001\,503 = 1.503 \times 10^{-3}$$
>
> a)   the first s.f. is 1
> b)   the third s.f. is 0

In each of the following numbers write down the significant figure specified in the bracket:

1. 36.2   (1st)   **6.** 5.083   (3rd)

2. 378.5   (3rd)   **7.** 34.807   (4th)

3. 0.0867   (2nd)   **8.** 0.076 03   (3rd)

4. 3.786   (3rd)   **9.** 54.06   (3rd)

5. 47 632   (2nd)   **10.** 5.7087   (4th)

EXERCISE 1k

> Give 32 685 correct to 1 s.f.
>
> (First write 32 685 in standard form.)
> $$32\,685 = 3.\!\mid\!2685 \times 10^4$$
>
> (As before, to correct to 1 s.f. we look at the second s.f.: if it is 5 or more we add one to the first s.f.; if it is less than 5 we leave the first s.f. alone.)
>
> So $3\!\mid\!2\,685 = 30\,000$ to 1 s.f.

Give the following numbers correct to 1 s.f.:

1. 59 727   **5.** 80 755   **9.** 667 505

2. 4164   **6.** 476   **10.** 908

3. 4 396 185   **7.** 51 488   **11.** 26

4. 586 359   **8.** 4099   **12.** 980

Give the following numbers correct to 2 s.f.:

13. 4673   **15.** 59 700   **17.** 6992

14. 57 341   **16.** 892 759   **18.** 9973

19.	72 601	**21.**	50 047	**23.**	476
20.	444	**22.**	53 908	**24.**	597

Give 0.021 94 correct to 3 s.f.

$$0.021\,94 = 2.19{\vert}4 \times 10^{-2}$$

(The fourth s.f. is 4 so we leave the third s.f. alone.)

So $\qquad 0.021\,9{\vert}4 = 0.0219$ to 3 s.f.

Give the following numbers correct to 3 s.f.:

25.	0.008 463	**30.**	0.007 854 7
26.	0.825 716	**31.**	7.5078
27.	5.8374	**32.**	369.649
28.	78.49	**33.**	0.989 624
29.	46.8451	**34.**	53.978

Give each of the following numbers correct to the number of significant figures indicated in the bracket.

35.	46.931 06	(2)		**40.**	4537	(1)
36.	0.006 845 03	(4)		**41.**	37.856 72	(3)
37.	576 335	(1)		**42.**	6973	(2)
38.	497	(2)		**43.**	0.070 865	(3)
39.	7.824 38	(3)		**44.**	0.067 34	(1)

Find $50 \div 8$ correct to 2 s.f.

(To give an answer correct to 2 s.f. we first work to 3 s.f.)

$$\frac{6.2{\vert}5}{8\,)\,50.00}$$

So $\qquad 50 \div 8 = 6.3$ to 2 s.f.

Give, correct to 2 s.f.

45. $20 \div 6$ **50.** $143 \div 5$

46. $10 \div 6$ **51.** $73 \div 3$

47. $25 \div 2$ **52.** $0.7 \div 3$

48. $53 \div 4$ **53.** $0.23 \div 9$

49. $125 \div 9$ **54.** $0.0013 \div 3$

ROUGH ESTIMATES

If you were asked to find 1.397×62.57 you could do it by long multiplication or you could use a calculator. Whichever method you choose, it is essential first to make a rough estimate of the answer. You will then know whether the actual answer you get is reasonable or not.

One way of estimating the answer to a calculation is to write each number correct to 1 significant figure.

So $1.|397 \times 6|2.57 \approx 1 \times 60 = 60$

EXERCISE 1I

Correct each number to 1 s.f. and hence give a rough answer to
a) 9.524×0.0837 b) $54.72 \div 0.761$

a) $9|.524 \times 0.08|37 \approx 10 \times 0.08 = 0.8$

b) $\dfrac{5|4.72}{0.7|61} \approx \dfrac{50}{0.8} = \dfrac{500}{8}$

$= 60$ (giving $500 \div 8$ to 1 s.f.)

Correct each number to 1 s.f. and hence give a rough answer to each of the following calculations:

1. 4.78×23.7 **6.** $82.8 \div 146$

2. 56.3×0.573 **7.** 0.632×0.845

3. $0.0674 \div 5.24$ **8.** 0.0062×574

4. 354.6×0.0475 **9.** $7.835 \div 6.493$

5. 576×256 **10.** 4736×729

11. 34.7×21

12. 8.63×0.523

13. $34.9 \div 15.8$

14. $0.47 \div 0.714$

15. $985 \div 57.2$

16. $0.0326 \div 12.4$

17. $0.007\,24 \times 0.783$

18. $3581 \div 45$

19. 1097×94

20. 45.07×0.0327

Correct each number to 1 s.f. and hence calculate
$$\dfrac{0.048 \times 3.275}{0.367} \text{ to 1 s.f.}$$

$$\frac{0.04\vert 8 \times 3.\vert 275}{0.3\vert 67} \approx \frac{0.05 \times 3}{0.4} = \frac{0.15}{0.4} = \frac{1.5}{4}$$
$$= 0.4 \quad \text{(to 1 s.f.)}$$

---

21. $\dfrac{3.87 \times 5.24}{2.13}$

22. $\dfrac{0.636 \times 2.63}{5.47}$

23. $\dfrac{21.78 \times 4.278}{7.96}$

24. $\dfrac{6.38 \times 0.185}{0.628}$

25. $\dfrac{43.8 \times 3.62}{4.72}$

26. $\dfrac{89.03 \times 0.079\,37}{5.92}$

27. $\dfrac{975 \times 0.636}{40.78}$

28. $\dfrac{8.735}{5.72 \times 5.94}$

29. $\dfrac{0.527}{6.41 \times 0.738}$

30. $\dfrac{57.8}{0.057 \times 6.93}$

CALCULATIONS: MULTIPLICATION AND DIVISION

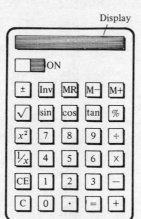

When you key in a number on your calculator it appears on the display. Check that the number on display is the number that you intended to enter.

EXERCISE 1m   First make a rough estimate of the answer. Then use your calculator to give the answer correct to 3 s.f.

1. 2.16×3.28
2. 2.63×2.87
3. 1.48×4.74
4. 4.035×2.116
5. 3.142×2.925

6. 6.053×1.274
7. 2.304×3.251
8. 8.426×1.086
9. $5.839 \div 3.618$
10. $6.834 \div 4.382$

11. $9.571 \div 2.518$
12. $5.393 \div 3.593$
13. $7.384 \div 2.51$
14. $4.931 \div 3.204$
15. $8.362 \div 5.823$

16. 23.4×56.7
17. 384×21.8
18. 45.8×143.7
19. $537.8 \div 34.6$
20. $45.35 \div 6.82$

21. 63.8×2.701
22. $40.3 \div 2.74$
23. $400 \div 35.7$
24. $(34.2)^2$
25. 5007×2.51

26. $5703 \div 154.8$
27. 39.03×49.94
28. $2000 \div 52.66$
29. $(36.8)^2$
30. $29\,006 \div 2.015$

31. 0.366×7.37
32. 0.0526×0.372
33. $6.924 \times 0.007\,93$
34. 0.638×825
35. 52×0.0895

36. 0.0826×0.582
37. 24.78×0.0724
38. $0.008\,35 \times 0.617$
39. 0.5824×6.813
40. $(0.74)^2$

41. $0.583 \div 4.82$
42. $0.628 \div 7.61$
43. $0.493 \div 1.253$
44. $0.518 \div 5.047$
45. $82.7 \div 593$

46. $89.5 \div 0.724$
47. $38.07 \div 0.682$
48. $5.71 \div 0.0623$
49. $7.045 \div 0.0378$
50. $6.888 \div 0.0072$

51.	$45.37 \div 0.925$	**56.**	$0.528 \div 0.0537$
52.	$8.41 \div 0.000\,748$	**57.**	$0.571 \div 0.824$
53.	$6.934 \div 0.0829$	**58.**	$0.0455 \div 0.0613$
54.	$0.824 \div 0.362$	**59.**	$0.006 \div 0.047\,03$
55.	$0.572 \div 0.851$	**60.**	$0.824 \div 0.000\,08$

61.	$5000 \div 0.789$	**68.**	$0.0467 \div 0.000\,074$
62.	$(0.078)^2$	**69.**	$(0.000\,31)^2$
63.	0.0608×573	**70.**	$\dfrac{54.9 \times 36.6}{0.406}$
64.	$(78.5)^3$	**71.**	$68.41 \div 392.9$
65.	$\dfrac{3.782 \times 0.467}{4.89}$	**72.**	$0.0482 \div 0.002\,89$
66.	$4.88 \times 0.004\,17$	**73.**	$(0.0527)^3$
67.	$0.9467 \div 7683$	**74.**	$\dfrac{0.857 \times 8.109}{0.5188}$

MIXED EXERCISES

EXERCISE 1n

1. Find the value of 4^{-2}.

2. Simplify $b^2 \div b^5$.

3. Find the value of $\dfrac{3^2 \times 3^3}{3^5}$.

4. Write $36\,400$ in standard form.

5. Write $0.005\,07$ in standard form.

6. Give $57\,934$ correct to 1 s.f.

7. Give $0.061\,374$ correct to 3 s.f.

8. Find 0.582×6.382, giving your answer correct to 3 s.f.

9. Find $45.823 \div 15.89$, giving your answer correct to 3 s.f.

EXERCISE 1p   **1.** Find the value of 6^3.

2. Write $\dfrac{2^4 \times 2^2}{2^8}$ as a single number in index form.

3. Find the value of $5^6 \div 5^7$.

4. Simplify $a^2 \times a^4 \times a$.

5. Write $650\,000\,000$ in standard form.

6. Give $45\,823$ correct to 2 s.f.

7. The organisers of a pop concert hope that, to the nearest thousand, $22\,000$ people will buy tickets. What is the minimum number of tickets that they hope to sell?

8. Find the value of $12.07 \div 0.008\,97$ giving your answer correct to 3 s.f.

9. Find the value of $(0.836)^2$ giving your answer correct to 3 s.f.

EXERCISE 1q   **1.** Find the value of $5^{-2} \times 5^3$.

2. Simplify $\dfrac{a^4}{a^3 \times a^2}$.

3. Find the value of $3^2 \times 3^4 \div 3^6$.

4. Write $0.005\,708$ in standard form.

5. Give 9764 correct to 1 s.f.

6. Give $0.050\,806$ correct to 3 s.f.

7. Correct to 1 significant figure, there are 70 matches in a box. What is the difference between the maximum and the minimum number of matches that could be in the box?

8. Find $0.0468 \div 0.004\,73$ giving your answer correct to 3 s.f.

9. Find $\dfrac{56.82 \times 0.714}{8.625}$ giving your answer correct to 3 s.f.

2 PROBABILITY

OUTCOMES OF EXPERIMENTS

If you throw an ordinary dice there are six possible scores that you can get. These are 1, 2, 3, 4, 5, or 6.

The act of throwing the dice is called an *experiment*.

The score that you get is called an *outcome* or an *event*.

The set {1, 2, 3, 4, 5, 6} is called the *set of all possible outcomes*.

EXERCISE 2a   How many possible outcomes are there for the following experiments? Write down the set of all possible outcomes in each case.

1. Tossing a 10 p coin. (Assume that it lands flat.)

2. Taking one disc from a bag containing 1 red, 1 blue and 1 yellow disc.

3. Choosing one number from the first ten integers. (An integer is a whole number.)

4. Taking one crayon from a box containing 1 red, 1 yellow, 1 blue, 1 brown, 1 black and 1 green crayon.

5. Taking one item from a bag containing 1 packet of chewing gum, 1 packet of boiled sweets and 1 bar of chocolate.

6. Taking one coin from a bag containing one 1p coin, one 10p coin, one 20p coin and one 50p coin.

7. Choosing one card from part of a pack of ordinary playing cards containing just the suit of clubs.

8. Choosing one letter from the vowels of the alphabet.

9. Choosing one number from the first 5 prime numbers.

10. Choosing an even number from the first 20 whole numbers.

PROBABILITY

If you throw an ordinary dice, what are the chances of getting a four? If you throw it fairly, it is reasonable to assume that you are as likely to throw any one score as any other, i.e. all outcomes are equally likely. As throwing a four is only 1 of the 6 equally likely outcomes you have a 1 in 6 chance of throwing a four.

"Odds" is another word in everyday language that is used to describe chances.

In mathematical language we use the word "probability" to describe chances. We say that the probability of throwing a four is $\frac{1}{6}$.
This can be written more briefly as

$$P(\text{throwing a four}) = \frac{1}{6}$$

We will now define exactly what we mean by "the probability that something happens".
If A stands for a particular event, the probability of A happening is written $P(A)$ where

$$P(A) = \frac{\text{the number of ways in which A can occur}}{\text{the } total \text{ number of equally likely outcomes}}$$

We can use this definition to work out, for example, the probability that if one card is drawn at random from a full pack of ordinary playing cards, it is the ace of spades.
(The phrase "at random" means that any one card is as likely to be picked as any other.)
There are 52 cards in a full pack, so there are 52 equally likely outcomes.
There is only one ace of spades, so there is only one way of drawing that card,

i.e. $$P(\text{ace of spades}) = \frac{1}{52}$$

EXERCISE 2b   In the following questions, assume that all possible outcomes are equally likely.

1. One letter is chosen at random from the letters in the word SALE. What is the probability that it is A?

2. What is the probability that a red pencil is chosen from a box containing 10 different coloured pencils?

3. What is the probability of choosing a prime number from the numbers 6, 7, 8, 9, 10?

4. What is the probability of picking the most expensive car from a range of six new cars in a showroom?

5. What is the probability of choosing an integer that is exactly divisible by 5 from the set {6, 7, 8, 9, 10, 11, 12}?

6. In a raffle 200 tickets are sold. If you have bought one ticket, what is the probability that you will win first prize?

7. One card is chosen at random from a pack of 52 ordinary playing cards. What is the probability that it is the ace of hearts?

8. What is the probability of choosing the colour blue from the colours of the rainbow?

9. A whole number is chosen from the first 15 whole numbers. What is the probability that it is exactly divisible both by 3 and by 4?

EXPERIMENTS WHERE AN EVENT CAN HAPPEN MORE THAN ONCE

If a card is picked at random from an ordinary pack of 52 playing cards, what is the probability that it is a five?

There are 4 fives in the pack, the five of spades, the five of hearts, the five of diamonds and the five of clubs.

That is, there are 4 ways in which a five can be picked.

Altogether there are 52 cards that are equally likely to be picked,

therefore $\qquad P(\text{picking a five}) = \frac{4}{52} = \frac{1}{13}$

Now consider a bag containing 3 white discs and 2 black discs.

If one disc is taken from the bag it can be black or white. But these are not equally likely events: there are three ways of choosing a white disc and two ways of choosing a black disc, so

$$P(\text{choosing a white disc}) = \frac{3}{5}$$

and $\qquad P(\text{choosing a black disc}) = \frac{2}{5}$

EXERCISE 2c

> A letter is chosen at random from the letters of the word DIFFICULT. How many ways are there of choosing the letter I? What is the probability that the letter I will be chosen?
>
> There are 2 ways of choosing the letter I and there are 9 letters in DIFFICULT.
>
> $$P(\text{choosing I}) = \frac{2}{9}$$

1. How many ways are there of choosing an even number from the first 10 whole numbers?

2. A prime number is picked at random from the set $\{4, 5, 6, 7, 8, 9, 10, 11\}$. How many ways are there of doing this?

3. A card is taken at random from an ordinary pack of 52 playing cards. How many ways are there of taking a black card?

4. An ordinary six-sided dice is thrown. How many ways are there of getting a score that is greater than 4?

5. A lucky dip contains 50 boxes, only 10 of which contain a prize, the rest being empty. How many ways are there of choosing a box that contains a prize?

6. A number is chosen at random from the first 10 integers. What is the probability that it is

   a) an even number   c) a prime number

   b) an odd number   d) exactly divisible by 3?

7. One card is drawn at random from an ordinary pack of 52 playing cards. What is the probability that it is

   a) an ace   c) a heart

   b) a red card   d) a picture card (include the aces)?

8. One letter is chosen at random from the word DIFFICULT. What is the probability that it is
    a) the letter F
    b) the letter D
    c) a vowel
    d) one of the first five letters of the alphabet?

9. An ordinary six-sided dice is thrown. What is the probability that the score is
    a) greater than 3   b) at least 5   c) less than 3?

10. A book of 150 pages has a picture on each of 20 pages. If one page is chosen at random, what is the probability that it has a picture on it?

11. One counter is picked at random from a bag containing 15 red counters, 5 white counters and 5 yellow counters. What is the probability that the counter removed is
    a) red   b) yellow   c) not red?

12. If you bought 10 raffle tickets and a total of 400 were sold, what is the probability that you win first prize?

13. A roulette wheel is spun. What is the probability that when it stops it will be pointing to
    a) an even number
    b) an odd number
    c) a number less than 10 excluding zero?
    (The numbers on a roulette wheel go from 0 to 35, and zero is neither an even number nor an odd number.)

14. One letter is chosen at random from the letters of the alphabet. What is the probability that it is a consonant?

15. A number is chosen at random from the set of two-digit numbers (i.e. the numbers from 10 to 99). What is the probability that it is exactly divisible both by 3 and by 4?

16. A bag of sweets contains 4 caramels, 3 fruit centres and 5 mints. If one sweet is taken out, what is the probability that it is
    a) a mint   b) a caramel   c) not a fruit centre?

CERTAINTY AND IMPOSSIBILITY

Consider a bag that contains 5 red discs only. If one disc is removed it is absolutely certain that it will be red. It is impossible to take a blue disc from that bag.

$$P(\text{disc is red}) = \frac{5}{5} = 1$$
$$P(\text{disc is blue}) = \frac{0}{5} = 0$$

In all cases

$$P(\text{an event that is certain}) = 1$$
$$P(\text{an event that is impossible}) = 0$$

Most events fall somewhere between the two, so

$$0 \leqslant P(\text{that an event happens}) \leqslant 1$$

EXERCISE 2d   Discuss the probability that the following events will happen. Try to class them as certain, impossible or somewhere in between.

1. You will swim the Atlantic Ocean.

2. You will weigh 80 kg.

3. You will be late home from school at least once this term.

4. You will grow to a height of 2 m.

5. The sun will not rise tomorrow.

6. You will run a mile in $3\frac{1}{2}$ minutes.

7. You will have a drink sometime today.

8. Newtown Football Club will win next year's F.A. Cup.

9. A card chosen from an ordinary pack of playing cards is either red or black.

10. A coin that is tossed lands on its edge.

11. Give some examples of events that are likely or unlikely to happen. For example: you will own a car; your home will burn down.

PROBABILITY THAT AN EVENT DOES NOT HAPPEN

If one card is drawn at random from an ordinary pack of playing cards, the probability that it is a club is given by

$$P(\text{a club}) = \frac{13}{52} = \frac{1}{4}$$

Now there are 39 cards that are not clubs so the probability that the card is not a club is given by

$$P(\text{not a club}) = \frac{39}{52} = \frac{3}{4}$$

i.e. $P(\text{not a club}) + P(\text{a club}) = \frac{3}{4} + \frac{1}{4} = 1$

Hence $P(\text{not a club}) = 1 - P(\text{a club})$

This relationship is true in any situation because

$$\begin{pmatrix} \text{The number of ways} \\ \text{in which an event, A,} \\ \text{can } not \text{ happen} \end{pmatrix} = \begin{pmatrix} \text{The total number of} \\ \text{possible outcomes} \end{pmatrix} - \begin{pmatrix} \text{The number of ways} \\ \text{in which A can} \\ \text{happen} \end{pmatrix}$$

i.e. $P(\text{A does not happen}) = 1 - P(\text{A does happen})$

"A does not happen" is shortened to $\bar{A}$, where $\bar{A}$ is read as "not A".

Therefore $\boxed{P(\bar{A}) = 1 - P(A)}$

EXERCISE 2e

A letter is chosen at random from the letters of the word PROBABILITY. What is the probability that it is not B?

Method 1:   There are 11 letters and 2 of them are Bs

$$\therefore \quad P(\text{letter is B}) = \frac{2}{11}$$

Hence $P(\text{letter is not B}) = 1 - \frac{2}{11}$

$$= \frac{9}{11}$$

Method 2: There are 11 letters and 9 of them are not Bs

$$\therefore \quad P(\text{letter is not B}) = \frac{9}{11}$$

1. A number is chosen at random from the first 20 whole numbers. What is the probability that it is not a prime number?

2. A card is drawn at random from an ordinary pack of playing cards. What is the probability that it is not a two?

3. One letter is chosen at random from the letters of the alphabet. What is the probability that it is not a vowel?

4. A box of 60 coloured crayons contains a mixture of colours, 10 of which are red. If one crayon is removed at random, what is the probability that it is not red?

5. A number is chosen at random from the first 10 whole numbers. What is the probability that it is not exactly divisible by 3?

6. One letter is chosen at random from the letters of the word ALPHABET. What is the probability that it is not a vowel?

7. In a raffle, 500 tickets are sold. If you buy 20 tickets, what is the probability that you will not win first prize?

8. If you throw an ordinary six-sided dice, what is the probability that you will not get a score of 5 or more?

9. There are 200 packets hidden in a lucky dip. Five packets contain £1 and the rest contain 1 p. What is the probability that you will not draw out a packet containing £1?

10. When an ordinary pack of playing cards is cut, what is the probability that the card showing is not a picture card? (The picture cards are the jacks, queens and kings.)

11. A letter is chosen at random from the letters of the word SUCCESSION. What is the probability that the letter is
a) N   b) S   c) a vowel   d) not S?

12. A card is drawn at random from an ordinary pack of playing cards. What is the probability that it is a) an ace b) a spade c) not a club   d) not a seven or an eight?

13. A bag contains a set of snooker balls (i.e. 15 red and 1 each of the following colours: white, yellow, green, brown, blue, pink and black). What is the probability that one ball removed at random is
a) red   b) not red   c) black   d) not red or white?

14. There are 60 cars in the station car park. Of the cars, 22 are British made, 24 are Japanese made and the rest are European but not British. What is the probability that the first car to leave is   a) Japanese   b) not British
c) European but not British   d) American?

POSSIBILITY SPACE FOR TWO EVENTS

Suppose a 2p coin and a 10p coin are tossed together. One possibility is that the 2p coin will land head up and that the 10p coin will also land head up.

If we use H for a head on the 2p coin and *H* for a head on the 10p coin, we can write this possibility more briefly as the ordered pair (H, *H*).

To list all the possibilities, an organized approach is necessary, otherwise we may miss some. We use a table called a *possibility space*. The possibilities for the 10p coin are written across the top and the possibilities for the 2p coin are written down the side:

10p coin

	H	*T*
H		
T		

2p coin

When both coins are tossed we can see all the combinations of heads and tails that are possible and then fill in the table.

10p coin

	H	*T*
H	(H, *H*)	(H, *T*)
T	(T, *H*)	(T, *T*)

2p coin

EXERCISE 2f   **1.** Two bags each contain 3 white counters and 2 black counters. One counter is removed at random from each bag. Copy and complete the following possibility space for the possible combinations of two counters.

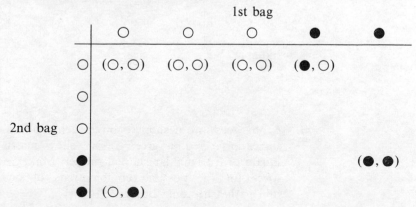

2. An ordinary six-sided dice is tossed and a 10p coin is tossed. Copy and complete the following possibility space.

Dice

	1	2	3	4	5	6
H		(H, 2)				
T				(T, 4)		

10p coin

3. One bag contains 2 red counters, 1 yellow counter and 1 blue counter. Another bag contains 2 yellow counters, 1 red counter and 1 blue counter. One counter is taken at random from each bag. Copy and complete the following possibility space.

1st bag

	R	R	Y	B
R		(R, R)		
Y				(B, Y)
Y				
B	(R, B)			

2nd bag

4. A top like the one in the diagram is spun twice. Copy and complete the possibility space.

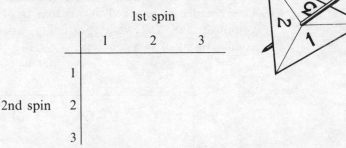

1st spin

	1	2	3
1			
2			
3			

2nd spin

5. A boy goes into a shop to buy a pencil and a rubber. He has a choice of a red, a green or a yellow pencil and a round, a square or a triangular shaped rubber. Make your own possibility space for the possible combinations of one pencil and one rubber that he could buy.

USING A POSSIBILITY SPACE

When there are several entries in a possibility space it can take a long time to fill in the ordered pairs. To save time we use a cross in place of each ordered pair. We can see which ordered pair a particular cross represents by looking at the edges of the table.

EXERCISE 2g

Two ordinary six-sided dice are tossed. Draw up a possibility space showing all the possible combinations in which the dice may land.

Use the possibility space to find the probability that a total score of at least 10 is obtained.

1st dice

		1	2	3	4	5	6
	1	×	×	×	×	×	×
	2	×	×	×	×	×	×
	3	×	×	×	×	×	×
2nd dice	4	×	×	×	×	×	⊗
	5	×	×	×	×	⊗	⊗
	6	×	×	×	⊗	⊗	⊗

(There are 36 entries in the table and 6 of these give a score of 10 or more.)

$$P(\text{score of at least 10}) = \frac{6}{36} = \frac{1}{6}$$

1. Use the possibility space in the example above to find the probability of getting a score of

a) 4 or less   b) 9   c) a double.

2. Use the possibility space for question 1 of Exercise 2f to find the probability that the two counters removed

a) are both black   b) contain at least one black.

3. Use the possibility space for question 2 of Exercise 2f to find the probability that the coin lands head up and the dice gives a score that is less than 3.

4. Use the possibility space for question 3 of Exercise 2f to find the probability that the two counters removed are

a) both blue

b) both red

c) one blue and one red

d) such that at least one is red.

5. A 5p coin and a 1p coin are tossed together. Make your own possibility space for the combinations in which they can land. Find the probability of getting two heads.

6. A six-sided dice has two of its faces blank and the other faces are numbered 1, 3, 4 and 6. This dice is tossed with an ordinary six-sided dice (faces numbered 1, 2, 3, 4, 5, 6). Make a possibility space for the ways in which the two dice can land and use it to find the probability of getting a total score of

a) 6   b) 10   c) 1   d) at least 6.

7. One bag of coins contains three 10p coins and two 50p coins. Another bag contains one 10p coin and one 50p coin. One coin is removed at random from each bag. Make a possibility space and use it to find the probability that a 50p coin is taken from each bag.

8. One bookshelf contains two storybooks and three textbooks. The next shelf holds three storybooks and one text book. Draw a possibility space showing the various ways in which you could pick up a pair of books, one from each shelf. Use this to find the probability that

a) both books are storybooks

b) both are textbooks.

9. The four aces and the four kings are removed from an ordinary pack of playing cards. One card is taken from the set of four aces and one card is taken from the set of four kings. Make a possibility space for the possible combinations of two cards and use it to find the probability that the two cards

a) are both black

b) are both spades

c) include at least one black card

d) are both of the same suit.

FINDING PROBABILITY BY EXPERIMENT

We have assumed that if you toss a coin it is equally likely to land head up or tail up so that P(a head) $= \frac{1}{2}$. Coins like this are called "fair" or "unbiased".

Most coins are likely to be unbiased but it is not necessarily true of all coins. A particular coin may be slightly bent or even deliberately biased so that there is not an equal chance of getting a head or a tail.

The only way to find out if a particular coin is unbiased is to toss it several times and count the number of times that it lands head up.

Then for that coin

$$P(\text{a head}) \approx \frac{\text{number of heads}}{\text{total number of tosses}}$$

The approximation gets nearer to the truth as the number of tosses gets larger.

EXERCISE 2h   Work with a partner or collect information from the whole class.

1. Toss a 2p coin 100 times and count the number of times it lands head up and the number of times it lands tail up.
   Use tally marks, in groups of five, to count as you toss.
   Find, approximately, the probability of getting a head with this coin.

2. Repeat question 1 with a 10p coin.

3. Repeat question 1 with the 2p coin that you used first but this time stick a small piece of plasticine on one side.

4. Choose two 2p coins and toss them both once. What do you think is the probability of getting two heads? Now toss the two coins 100 times and count the number of times that both coins land head up together. Use tally marks to count as you go: you will need to keep two tallies, one to count the total number of tosses and one to count the number of times you get two heads. Use your results to find approximately the probability of getting two heads.

5. Take an ordinary pack of playing cards and keep them well shuffled. If the pack is cut, what do you think is the probability of getting a red card? Cut the pack 100 times and keep count, using tally marks as before, of the number of times that you get a red card. Now find an approximate value for the probability of getting a red card.

6. Using the pack of cards again, what do you think is the probability of getting a spade? Now find this probability by experiment.

7. Use an ordinary six-sided dice. Toss it 25 times and keep count of the number of times that you get a six. Use your results to find an approximate value for the probability of getting a six. Now toss the dice another 25 times and add the results to the last set. Use these to find again the probability of getting a six. Now do another 25 tosses and add the results to the last two sets to find another value for the probability. Carry on doing this in groups of 25 tosses until you have done 200 tosses altogether.

   You know that the probability of getting a six is $\frac{1}{6}$. Now look at the sequence of results obtained from your experiment. What do you notice? (It is easier to compare your results if you use your calculator to change the fraction into decimals correct to 2 d.p.)

8. Remove all the diamonds from an ordinary pack of playing cards. Shuffle the remaining cards well and then cut the pack. What do you think is the probability of getting a black card? Shuffle and cut the pack 100 times and use the results to find approximately the probability of cutting a black card.

9. Take two ordinary six-sided dice and toss them both. What do you think is the probability of getting two 6s? Find this probability by experiment: you will need to do about 200 tosses to get a reasonable answer.

3 CONSTRUCTIONS

ANGLES AND TRIANGLES

Reminder:

Vertically opposite angles are equal.

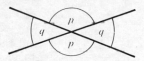

Angles at a point add up to 360°.

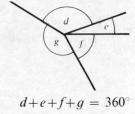

$$d+e+f+g = 360°$$

Angles on a straight line add up to 180°.

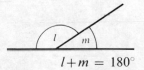

$$l+m = 180°$$

The sum of the three angles in any triangle is 180°.

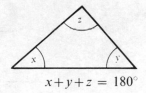

$$x+y+z = 180°$$

The sum of the four angles in any quadrilateral is 360°.

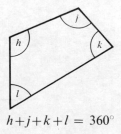

$$h+j+k+l = 360°$$

33

An equilateral triangle has all three sides the same length and each of the three angles is 60°.

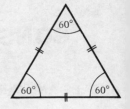

An isosceles triangle has two equal sides and the two angles at the base of the equal sides are equal.

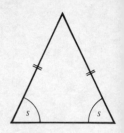

When a transversal cuts a pair of parallel lines:

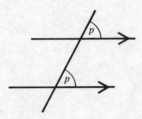

the corresponding angles are equal

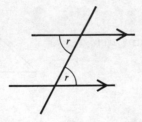

the alternate angles are equal

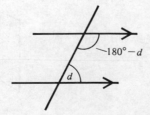

the interior angles are supplementary (add up to 180°)

EXERCISE 3a Find the sizes of the marked angles. If two angles are marked with the same letter they are the same size.

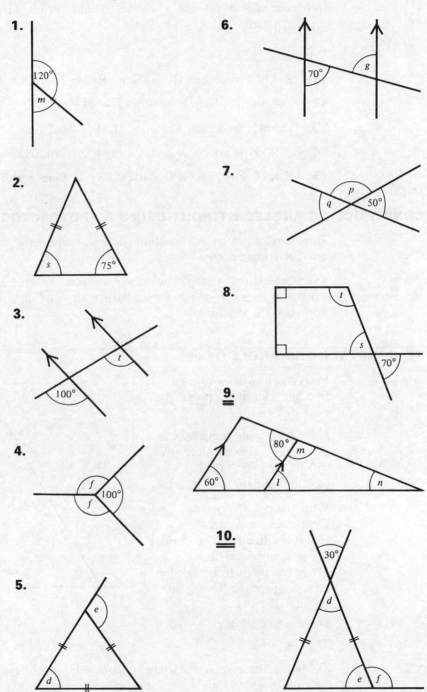

In Book 1 you learnt how to construct triangles. Before you start a construction, remember to make a rough sketch and to put all the information that you are given on to that sketch. Then decide which method to use.

Construct

11. $\triangle ABC$ in which $AB = 5\,cm$, $BC = 7\,cm$ and $AC = 6\,cm$

12. $\triangle PQR$ in which $\hat{P} = 60°$, $\hat{Q} = 40°$ and $PQ = 8\,cm$

13. $\triangle LMN$ in which $\hat{M} = 45°$, $LM = 7\,cm$ and $MN = 8\,cm$

14. $\triangle XYZ$ in which $\hat{X} = 100°$, $\hat{Y} = 20°$ and $XY = 5\,cm$

15. $\triangle RST$ in which $RS = 10\,cm$, $ST = 6\,cm$ and $RT = 7\,cm$

CONSTRUCTING ANGLES WITHOUT USING A PROTRACTOR

Some angles can be made without using a protractor: one such angle is 60°.

Every equilateral triangle, whatever its size, has three angles of 60°. To make an angle of 60° we construct an equilateral triangle but do not draw the third side.

TO CONSTRUCT AN ANGLE OF 60°

Start by drawing a straight line and marking a point, A, near one end.

Next open your compasses to a radius of 4 cm or more (this will be the length of the sides of your equilateral triangle).

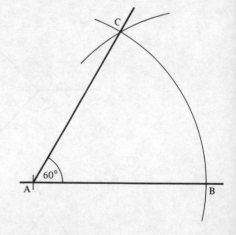

With the point of your compasses on A, draw an arc to cut the line at B, continuing the arc above the line.
Move the point to B and draw an arc above the line to cut the first arc at C.

Draw a line through A and C.

Then $\hat{A}$ is 60°.

$\triangle ABC$ is the equilateral triangle so *be careful not to alter the radius on your compasses during this construction.*

BISECTING ANGLES

Bisect means "cut exactly in half".

The construction for bisecting an angle makes use of the fact that, in an isosceles triangle the line of symmetry cuts $\hat{A}$ in half.

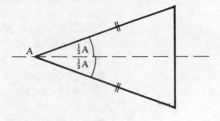

To bisect $\hat{A}$, open your compasses to a radius of about 6 cm.

With the point on A, draw an arc to cut both arms of $\hat{A}$ at B and C. (If we joined BC, △ABC would be isosceles.)

With the point on B, draw an arc between the arms of $\hat{A}$.

Move the point to C (being careful not to change the radius) and draw an arc to cut the other arc at D.

Join AD.

The line AD then bisects $\hat{A}$.

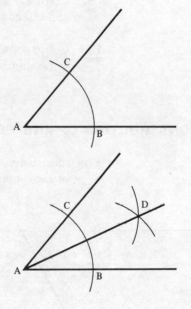

EXERCISE 3b

1. Construct an angle of 60°.

2. Draw an angle of about 50°. Bisect this angle. Measure both halves of your angle.

3. Construct an angle of 60°. Now bisect this angle. What size should each new angle be? Measure both of them.

4. Use what you learned from the last question to construct an angle of 30°.

5. Draw a straight line and mark a point A near the middle.

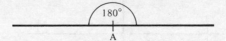

You now have an angle of 180° at A.

6. Draw an angle of 180° and then bisect it. What is the size of each new angle? Measure each of them.

7. Use what you learned from the last question to construct an angle of 90°.

8. Construct an angle of 45°. (Begin by constructing an angle of 90° and then bisect it.)

9. Construct an angle of 15°. (Start by constructing an angle of 60° and bisect as often as necessary.)

10. Construct an angle of 22.5°. (Start with 90° and bisect as often as necessary.)

CONSTRUCTION OF ANGLES OF 60°, 30°, 90°, 45°

You constructed these angles in the last exercise. Here is a summary of these constructions.

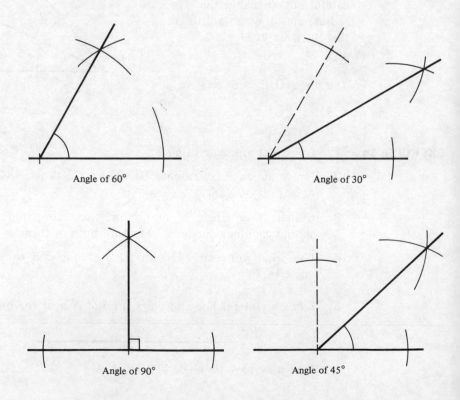

Angle of 60°   Angle of 30°

Angle of 90°   Angle of 45°

EXERCISE 3c   Construct the following figures using only a ruler and a pair of compasses:

1.

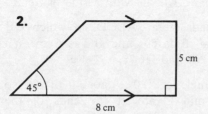

2.

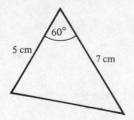

3.

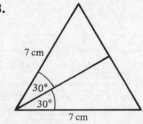

4.

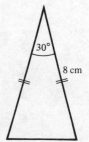

5.

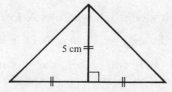

6.

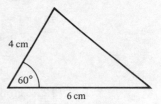

7.

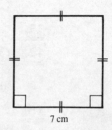

8.

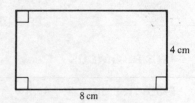

9.

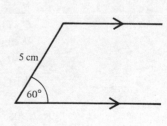

10.

For questions 11 to 15, draw a rough sketch before starting the construction.

11. Draw a line, AB, 12 cm long. Construct an angle of 60° at A. Construct an angle of 30° at B. Label with C the point where the arms of $\widehat{A}$ and $\widehat{B}$ cross. What size should $\widehat{C}$ be? Measure $\widehat{C}$ as a check on your construction.

12. Construct a triangle, ABC, in which AB is 10 cm long, $\widehat{A}$ is 90° and AC is 10 cm long. What size should $\widehat{C}$ and $\widehat{B}$ be? Measure $\widehat{C}$ and $\widehat{B}$ as a check.

13. Construct a square, ABCD, with a side of 6 cm.

14. Construct a quadrilateral, ABCD, in which AB is 12 cm, $\widehat{A}$ is 60°, AD is 6 cm, $\widehat{B}$ is 60° and BC is 6 cm. What can you say about the lines AB and DC?

15. Construct an angle of 120°. Label it BAC (so that A is the vertex and B and C are at the ends of the arms). At C, construct an angle of 60° so that $\widehat{C}$ and $\widehat{A}$ are on the same side of AC. You have constructed a pair of parallel lines; mark them and devise your own check.

THE RHOMBUS

EXERCISE 3d **1.** Draw a line 12 cm long across your page. Label the ends A and C. Open your compasses to a radius of 9 cm. With the point on A, draw an arc above AC and another arc below AC. Keeping the same radius, move the point of your compasses to C. Draw arcs above and below AC to cut the first pair of arcs. Where the arcs intersect (i.e. cross) label the points B and D. Join A to B, B to C, C to D and D to A.

ABCD is called a rhombus.

Questions 2 to 9 refer to the figure that you have constructed in question 1.

2. Without measuring them, what can you say about the lengths of AB, BC, CD and DA?

3. ABCD has two lines of symmetry. Name them.

4. If ABCD is folded along BD, where is A in relation to C?

5. If ABCD is folded along AC, where is D in relation to B?

6. Where AC and BD cut, label the point E. With ABCD unfolded, where is E in relation to A and C?

7. Where is E in relation to B and D?

8. If ABCD is folded first along BD and then folded again along AE, what is the size of the angle at E?

9. With ABCD unfolded, what are the sizes of the four angles at E?

PROPERTIES OF THE DIAGONALS OF A RHOMBUS

From the last exercise you should be convinced that

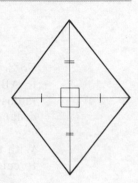

> the diagonals of a rhombus bisect each other at right angles.

These properties form the basis of the next two constructions.

CONSTRUCTION TO BISECT A LINE

To bisect a line we have to find the midpoint of that line. To do this we construct a rhombus with the given line as one diagonal, but we do not join the sides of the rhombus.

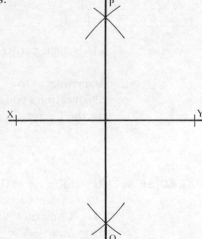

To bisect XY, open your compasses to a radius that is about $\frac{3}{4}$ of the length of XY.

With the point on X, draw arcs above and below XY.

Move the point to Y (being careful not to change the radius) and draw arcs to cut the first pair at P and Q.

Join PQ.

The point where PQ cuts XY is the midpoint of XY.

(XPYQ is a rhombus since the same radius is used to draw all the arcs, i.e. XP = YP = YQ = XQ. PQ and XY are the diagonals of the rhombus so PQ bisects XY.)

Note. When you are going to bisect a line, draw it so that there is plenty of space for the arcs above *and* below the line.

DROPPING A PERPENDICULAR FROM A POINT TO A LINE ─────

If you are told to drop a perpendicular from a point, C, to a line, AB, this means that you have to draw a line through C which is at right angles to the line AB.

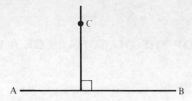

To drop a perpendicular from C to AB, open your compasses to a radius that is about $1\frac{1}{2}$ times the distance of C from AB.

With the point on C, draw arcs to cut the line AB at P and Q.

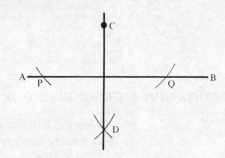

Move the point to P and draw an arc on the other side of AB. Move the point to Q and draw an arc to cut the last arc at D.

Join CD.

CD is then perpendicular to AB.

Remember to keep the radius unchanged throughout this construction: you then have a rhombus, PCQD, of which CD and PQ are the diagonals.

EXERCISE 3e   Remember to make a rough sketch before you start each construction.

1. Construct a triangle ABC, in which AB = 6 cm, BC = 8 cm and CA = 10 cm. Using a ruler and compasses only, drop a perpendicular from B to AC.

2. Construct a triangle ABC, in which AB = 8 cm, AC = 10 cm and CB = 9 cm. Drop a perpendicular from C to AB.

3. Construct a triangle XYZ, in which XY = 12 cm, XZ = 5 cm and YZ = 9 cm. Drop a perpendicular from Z to XY.

4. Construct the isosceles triangle LMN in which LM = 6 cm, LN = MN = 8 cm. Construct the perpendicular bisector of the side LM. Explain why this line is a line of symmetry of △LMN.

5. Construct the isosceles triangle PQR, in which PQ = 5 cm, PR = RQ = 7 cm. Construct the perpendicular bisector of the side PR. This line is not a line of symmetry of △PQR; why not?

6. The figure on the right is a circle whose centre is C, with a line, AB, drawn across the circle.
(AB is called a *chord*.)
This figure has one line of symmetry which is not shown. Make a rough sketch of the figure and mark the line of symmetry. Explain what the line of symmetry is in relation to AB.

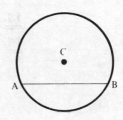

7. Draw a circle of radius 6 cm and mark the centre, C. Draw a chord, AB, about 9 cm long. (Your drawing will look like the one in question 6.) Construct the line of symmetry.

8. Construct a triangle ABC, in which AB = 8 cm, BC = 10 cm and AC = 9 cm. Construct the perpendicular bisector of AB. Construct the perpendicular bisector of BC. Where these two perpendicular bisectors intersect (i.e. cross), mark G. With the point of your compasses on G and with a radius equal to the length of GA, draw a circle.
This circle should pass through B and C, and it is called the *circumcircle* of △ABC.

9. Repeat question 8 with a triangle of your own.

10. Construct a square ABCD, such that its sides are 5 cm long. Construct the perpendicular bisector of AB and the perpendicular bisector of BC. Label with E the point where the perpendicular bisectors cross. With the point of your compasses on E and the radius equal to the distance from E to A, draw a circle.
This circle should pass through all four corners of the square. It is called the circumcircle of ABCD.

11. Construct a triangle ABC, in which AB = 10 cm, AC = 8 cm and BC = 12 cm. Construct the bisector of Â and the bisector of B̂. Where these two angle bisectors cross, mark E. Drop the perpendicular from E to AB. Label G, the point where this perpendicular meets AB. With the point of your compasses on E and the radius equal to EG, draw a circle.
This circle should touch all three sides of △ABC and it is called the *incircle* of △ABC.

12. Repeat question 11 with the equilateral triangle ABC, with sides that are 10 cm long.

13. Repeat question 11 with a triangle of your own.

14. Construct a square ABCD, of side 8 cm. Construct the incircle (i.e. the circle that *touches* all four sides of the square) of ABCD. First decide how you are going to find the centre of the circle.

MAKING SOLIDS

To make a solid object from a sheet of flat paper you need to construct a *net*: this is the shape that has to be cut out, folded and stuck together to make the solid. A net should be drawn as accurately as possible, otherwise you will find that the edges will not fit together properly.

EXERCISE 3f Each solid in this exercise has flat faces (called *plane* faces) and is called a polyhedron.
"Poly" is a prefix used quite often; it means "many".

1. The Tetrahedron
The net consists of four equilateral triangles. Construct the net accurately making the sides of each triangle 6 cm long. Start by drawing one triangle of side 12 cm; mark the midpoints of the sides and join them up. Draw flaps on the edges shown.

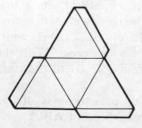

Cut out the net. Score the solid lines (use a ruler and ballpoint pen — an empty one is best) and fold the outer triangles up so that their vertices meet. Use the flaps to stick the edges together.

This solid is called a *regular* tetrahedron. A regular solid is one in which all the faces are identical. These make good Christmas tree decorations if painted or if made out of foil-covered paper.

2. Octahedron

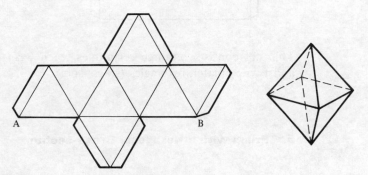

This net consists of equilateral triangles: make the sides of each triangle 4 cm long, and start by making AB 12 cm long. Is this octahedron a regular solid?

3. Square-based Pyramid

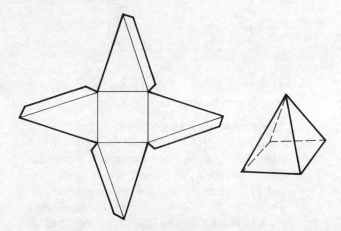

This net consists of a square with an isosceles triangle on each side of the square. Make the sides of the square 6 cm and the equal sides of the triangles 10 cm long. Is this a regular solid?

4. Prism with Triangular Cross-section

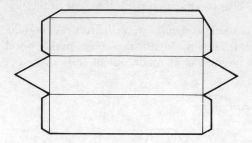 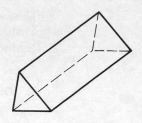

This net consists of three rectangles, each 8 cm long and 4 cm wide, and two equilateral triangles (sides 4 cm).

5. Prism with a Hexagonal Cross-section

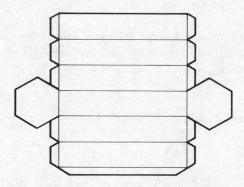 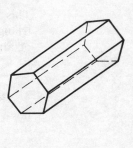

This net consists of six rectangles, each 8 cm long and 4 cm wide, and two hexagons each of side 4 cm.

The easiest way to construct a hexagon is to draw a circle of radius 4 cm and mark a point, A, on the circumference. With the point of the compasses on A and the radius kept at 4 cm, draw an arc to cut the circle at B. Move the point to B and repeat. Continue until you have reached A again. Join up the marks on the circle.

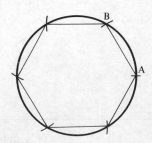

Cut out the hexagon and use it to draw round when constructing the net of the prism.

6. Eight-pointed Star (Stella Octangula)

This model needs time and patience. If you have both it is wort~~ ~~
effort!

It consists of a regular octahedron (see question 2) with a regular
tetrahedron (see question 1) stuck on each face.
You will need 8 tetrahedra. In all the nets make the triangles have
sides of length 4 cm.

DUCING PERCENTAGES

RCENTAGES AS FRACTIONS

'er cent" means per hundred, i.e. if 60 per cent of the workers in a factory are women it means that 60 out of every 100 workers are women. If there are 700 workers in the factory, $60 \times 7 = 420$ are women, while if there are 1200 workers, $60 \times 12 = 720$ are women.

In mathematics we are always looking for shorter ways of writing statements and especially for symbols to stand for words. The symbol that means "per cent" is %, i.e. 60 per cent and 60% have exactly the same meaning.

60 per cent means 60 per hundred and this can be written as the fraction $\frac{60}{100}$ (or $\frac{3}{5}$)
i.e. 60% of a quantity is exactly the same as $\frac{60}{100}$ (or $\frac{3}{5}$) of that quantity.

If there are 800 cars in a car park and 60% of them are British, then $\frac{60}{100}$ of the cars are British,

i.e.   the number of British cars is $\frac{60}{100} \times 800 = 480$

EXERCISE 4a

Express   a) 40% b) $22\frac{1}{2}\%$ as fractions in their lowest terms.

a) $40\% = \frac{40}{100} = \frac{2}{5}$

b) $22\frac{1}{2}\% = \frac{45}{2 \times 100} = \frac{9}{40}$

Express as fractions in their lowest terms:

1. 20%	**8.** 50%	**15.** 70%	**22.** 95%
2. 45%	**9.** 65%	**16.** 75%	**23.** 15%
3. 25%	**10.** 56%	**17.** 48%	**24.** 8%
4. 72%	**11.** 37%	**18.** 69%	**25.** 82%
5. $33\frac{1}{3}\%$	**12.** $66\frac{2}{3}\%$	**19.** $37\frac{1}{2}\%$	**26.** $87\frac{1}{2}\%$
6. $12\frac{1}{2}\%$	**13.** $62\frac{1}{2}\%$	**20.** $5\frac{1}{3}\%$	**27.** $6\frac{1}{4}\%$
7. $2\frac{1}{2}\%$	**14.** 125%	**21.** $17\frac{1}{2}\%$	**28.** 150%

Express a) 54% b) $6\frac{1}{2}\%$ c) $27\frac{1}{3}\%$ as decimals.

$$\text{a)} \quad 54\% = \frac{54}{100} = 0.54$$

$$\text{b)} \quad 6\frac{1}{2}\% = \frac{6.5}{100} = 0.065$$

$$\text{c)} \quad 27\frac{1}{3}\% = \frac{82}{3 \times 100} = 0.273 \text{ to 3 s.f.}$$

Express the following percentages as decimals, giving your answers correct to 3 s.f. where necessary:

29.	47%	**34.**	58%	**39.**	92%	**44.**	8%
30.	12%	**35.**	30%	**40.**	65%	**45.**	3%
31.	$5\frac{1}{2}\%$	**36.**	$62\frac{1}{4}\%$	**41.**	120%	**46.**	180%
32.	145%	**37.**	350%	**42.**	231%	**47.**	$5\frac{1}{3}\%$
33.	$58\frac{1}{3}\%$	**38.**	$48\frac{2}{3}\%$	**43.**	$85\frac{2}{3}\%$	**48.**	$54\frac{1}{7}\%$

EXPRESSING FRACTIONS AS PERCENTAGES

If $\frac{4}{5}$ of the pupils in a school have been away for a holiday, it means that 80 in every 100 have been on holiday,
i.e. $\frac{4}{5}$ is the same as 80%.

A fraction may be converted into a percentage by multiplying that fraction by 100%. This does not alter its value, since 100% is 1.

EXERCISE 4b

Express $\frac{7}{20}$ as a percentage.

$$\frac{7}{20} = \frac{7}{20_1} \times \overset{5}{\cancel{100}}\% = 35\%$$

Express the following fractions as percentages, giving your answers correct to 1 decimal place where necessary:

1.	$\frac{1}{2}$	**6.**	$\frac{1}{4}$	**11.**	$\frac{3}{4}$	**16.**	$\frac{3}{5}$
2.	$\frac{7}{10}$	**7.**	$\frac{3}{20}$	**12.**	$\frac{9}{20}$	**17.**	$\frac{7}{20}$
3.	$\frac{13}{20}$	**8.**	$\frac{4}{25}$	**13.**	$\frac{7}{5}$	**18.**	$\frac{31}{25}$
4.	$\frac{1}{3}$	**9.**	$\frac{3}{8}$	**14.**	$\frac{5}{8}$	**19.**	$\frac{7}{8}$
5.	$\frac{21}{40}$	**10.**	$\frac{23}{60}$	**15.**	$\frac{8}{3}$	**20.**	$\frac{8}{5}$

Express a) 0.7   b) 1.24   as percentages.

a) $0.7 = 0.7 \times 100\% = 70\%$

b) $1.24 = 1.24 \times 100\% = 124\%$

Express the following decimals as percentages:

21.	0.5	**26.**	0.9	**31.**	0.25	**36.**	0.36
22.	0.22	**27.**	0.04	**32.**	0.74	**37.**	0.16
23.	0.83	**28.**	0.55	**33.**	1.25	**38.**	1.39
24.	1.72	**29.**	2.64	**34.**	3.41	**39.**	6.35
25.	0.625	**30.**	0.845	**35.**	0.075	**40.**	0.1825

EXERCISE 4c

1. Express as fractions in their lowest terms:

a) 30% b) 85% c) $42\frac{1}{2}\%$ d) $5\frac{1}{4}\%$

2. Express as decimals:

a) 44% b) 68% c) 170% d) $16\frac{1}{2}\%$

3. Express as percentages:

a) $\frac{2}{5}$ b) $\frac{17}{20}$ c) $\frac{1}{8}$ d) $\frac{17}{15}$

4. Express as percentages:

a) 0.2   b) 0.62   c) 0.845   d) 1.78

Copy and complete the following table:

	Fraction	Percentage	Decimal
	$\frac{3}{4}$	75%	0.75
5.	$\frac{4}{5}$		
6.		60%	
7.			0.7
8.	$\frac{11}{20}$		
9.		44%	
10.			0.32

PROBLEMS

Suppose that in the town of Doxton 55 families in every 100 own a car. We can deduce from this that 45 in every 100 families do not. Since every family either owns a car or does not own a car, if we are given one percentage we can deduce the other.

EXERCISE 4d

> If 56% of homes have a telephone, what percentage do not?
>
> All homes either have, or do not have, a telephone.
>
> If 56% have a telephone, then $(100-56)\%$ do not,
>
> i.e. 44% do not.

1. If 48% of the pupils in a school are girls, what percentage are boys?

2. If 87% of households have a television set, what percentage do not?

3. In the fourth year, 64% of the pupils do not study chemistry. What percentage study chemistry?

4. In a box of oranges, 8% are bad. What percentage are good?

5. Twelve per cent of the cars that come to an MOT testing station fail to pass first time. What percentage pass first time?

6. A hockey team won 62% of their matches and drew 26% of them. What percentage did they lose?

7. A rugby team drew 12% of their matches and lost 45% of them. What percentage did they win?

8. Deductions from a youth's wage were: income tax 18%, other deductions 14%. What percentage did he keep?

9. In an election, 40% of the electorate voted for Mrs Long, 32% for Mr Singhe and the remainder voted for Miss Berry. What percentage voted for Miss Berry if there were only three candidates and 8% of the electorate failed to vote?

10. In a school, 36% of the pupils study French and 38% study German. If 12% study both languages, what percentage do not study either?

11. 85% of the first year pupils in a school study craft and 72% study photography. If 60% study both subjects, what percentage study neither?

12. A concert is attended by 1200 people. If 42% are adult females and 37% are adult males, how many children attended?

13. The attendance at an athletics meeting is 14 000. If 68% are men and boys and 22% are women, how many are girls?

14. In a book, 98% of the pages contain text, diagrams or both. If 88% of the pages contain text and 32% contain diagrams, what percentage contain

a) neither text nor diagrams

b) only diagrams

c) only text

d) both text and diagrams?

EXPRESSING ONE QUANTITY AS A PERCENTAGE OF ANOTHER

If we wish to find 4 as a percentage of 20, we know that 4 is $\frac{4}{20}$ of 20

and $$\frac{4}{20} = \frac{4}{20} \times 100\%$$

i.e. 4 as a percentage of 20 is

$$\frac{4}{20} \times 100\% = 20\%$$

To express one quantity as a percentage of another, we divide the first quantity by the second and multiply this fraction by 100%.

EXERCISE 4e

Express 20 cm as a percentage of 3 m.

(First express .3 m in centimetres to bring both quantities to the same unit.)

$$3 \,m = 3 \times 100 \,cm = 300 \,cm$$

Then the first quantity as a percentage of the second quantity is

$$\frac{20}{300} \times 100\% = 6\frac{2}{3}\%$$

Express the first quantity as a percentage of the second:

1. 3, 12
2. 30 cm, 50 cm
3. 3 m, 9 m
4. 4 in, 12 in

5. 15, 20
6. 24 cm, 40 cm
7. 60 cm, 4 m
8. 10 ft, 40 ft

9. 5, 50
10. 2 cm, 10 cm
11. 600 m, 2 km
12. $3\frac{1}{2}$ yd, 7 yd

13. 40, 20
14. 35 m, 56 m
15. 50 cm, 5 m
16. 8 in, 12 in

17. 20 m², 80 m²
18. 75 cm², 200 cm²
19. 25 cm², 125 cm²
20. 4 litres, 10 litres
21. 3 pints, 5 pints

22. 200 mm², 800 mm²
23. 198 mm², 275 mm²
24. 50 m², 15 m²
25. 3.6 t, 5 t
26. 33.6 g, 80 g

27. 1200 g, 3 kg
28. 3.64 kg, 5.6 kg
29. 28 cm, 1.2 m
30. 74 p, £1.11
31. 37 mm, 148 cm

32. 900 g, 2.5 kg
33. 45 p, £1.35
34. 98 mm, 2.45 m
35. 4 mm, 3 cm
36. 84 g, 3.36 kg

37. 46 cm², 1 m²
38. 10 cm², 200 mm²
39. 39 ft², 60 ft²
40. 72 in², 2 ft²
41. 0.1 m², 25 000 mm²

42. 100 cm³, 1 litre
43. 25 000 cm³, 1 m³
44. 6 pints, 3 gallons
45. £5.40, 81 p
46. 0.01 m³, 125 000 cm³

FINDING A PERCENTAGE OF A QUANTITY ──────────────

EXERCISE 4f

Find the value of   a) 12% of 450   b) $7\frac{1}{3}\%$ of 3.75 m

a) 12% of $450 = \frac{12}{100} \times 450 = 54$

b) $7\frac{1}{3}\%$ of $3.75\,\text{m} = 7\frac{1}{3}\%$ of $375\,\text{cm}$

$= \frac{22}{3 \times 100} \times 375\,\text{cm}$

$= 27.5\,\text{cm}$

Find the value of:

1. 40% of 120

2. 12% of 800 g

3. 74% of 75 cm

4. 44% of 650 km

5. 8% of £2

6. 77% of 4 kg

7. 70% of 360

8. 86% of 1150 g

9. 55% of 8.6 m

10. 96% of 215 cm^2

11. 63% of 4 m

12. 96% of 15 m^2

13. 45% of 740

14. 33% of 600 kg

15. 6% of 24 m

16. 15% of £10

17. 17% of 2 km

18. 32% of 5 litres

19. 30% of £250

20. 66% of 300 m

21. $33\frac{1}{3}\%$ of 270 g

22. $5\frac{1}{4}\%$ of 56 mm

23. $37\frac{1}{2}\%$ of 48 cm

24. $22\frac{1}{2}\%$ of 40 m^2

25. $66\frac{2}{3}\%$ of 480 m^2

26. $32\frac{1}{7}\%$ of 140 km

27. $62\frac{1}{2}\%$ of 8 km

28. $74\frac{1}{2}\%$ of 200 cm^2

29. $33\frac{1}{3}\%$ of 42 p

30. $82\frac{1}{5}\%$ of £65

31. 12% of £4

32. $7\frac{1}{2}\%$ of 80 g

33. $2\frac{1}{3}\%$ of 90 m

34. $16\frac{2}{3}\%$ of £60

35. $3\frac{1}{8}\%$ of 64 kg

36. $87\frac{1}{2}\%$ of 16 mm

PROBLEMS

EXERCISE 4g

> In the second year, 287 of the 350 pupils study geography. What percentage study geography?
>
> $$\text{Percentage studying geography} = \frac{287}{350} \times 100\%$$
> $$= 82\%$$

1. There are 60 boys in the third year, 24 of whom study chemistry. What percentage of third year boys study chemistry?

2. In a history test, Pauline scored 28 out of a possible 40. What was her percentage mark?

3. Out of 20 cars tested in one day by an MOT testing station, 4 of them failed. What percentage failed?

4. There are 60 photographs in a book, 12 of which are coloured. What is the percentage of coloured photographs?

5. Forty-two of the 60 choristers in a choir wear spectacles. What percentage do not?

6. Each week a boy saves £3 of the £12 he earns. What percentage does he spend?

7. A secretary takes 56 letters to the post office for posting. 14 are first class and the remainder are second class. What percentage go second class?

8. Judy obtained 80 marks out of a possible 120 in her end of term maths examination. What was her percentage mark?

9. Jane's gross wage is £120 per week, but her "take home" pay is only £78. What percentage is this of her gross wage?

10. If 8% of a crowd of 24 500 at a football match were females, how many females attended?

> If 54% of the 1800 pupils in a school are boys, how many girls are there in the school?
>
> $$\text{Number of boys} = \frac{54}{100} \times 1800$$
> $$= 972$$
> $$\text{Number of girls} = 1800 - 972$$
> $$= 828$$

11. In a garage, 16 of the 30 cars which are for sale are second hand. What percentage of the cars are

a) new   b) second hand?

12. There are 80 houses in my street and 65% of them have a telephone. How many houses

a) have a telephone   b) do not have a telephone?

13. In my class there are 30 pupils and 40% of them have a bicycle. How many pupils

a) have a bicycle   b) do not have a bicycle?

14. Yesterday, of the 240 flights leaving London Airport, 15% were bound for North America. How many of these flights

a) flew to North America   b) did not fly to North America?

15. In a particular year, 64% of the 16000 Jewish immigrants into Israel came from Eastern Europe. How many of the immigrants did not come from Eastern Europe?

16. There are 120 shops in the High Street, 35% of which sell food. How many High Street shops do not sell food?

17. Last year the amount I paid in rates on my house was £520. This year my rates will increase by 12%. Find the increase.

18. A mathematics book has 320 pages, 40% of which are on algebra, 25% on geometry and the remainder on arithmetic. How many pages of arithmetic are there?

MIXED EXERCISES

EXERCISE 4h   **1.** Express as a fraction in its lowest terms

a) 40%   b) 54%   c) $27\frac{1}{2}\%$

2. Express as a percentage

a) $\frac{3}{5}$ b) 0.78   c) 0.125

3. Express 2 m as a percentage of 25 m.

4. Express 25 p as a percentage of £2.

5. Find 45% of 120 m.

6. If 3% of telephone calls are connected to the wrong number, what percentage of calls are connected to the correct number?

EXERCISE 4i

1. Express 36%

a) as a vulgar fraction in its lowest terms   b) as a decimal.

2. Express as a percentage, giving your answer correct to 3 s.f. if necessary,

a) $\frac{5}{8}$ b) $1\frac{1}{3}$ c) 2.5

3. Express 250 g as a percentage of 2 kg.

4. Find 85% of 340 m².

5. The cost of insuring a car in central London is about 12% of its value. Find the cost of insuring a car valued at £7000.

EXERCISE 4j

1. Find the first quantity as a percentage of the second quantity:

a) 10 m, 80 m   b) 75 p, £2   c) 150 cm, 3 m

2. Express as a percentage, giving your answer correct to 3 s.f. where necessary,

a) $\frac{2}{7}$ b) 0.279   c) $1\frac{2}{9}$

3. Express $12\frac{1}{2}$% as

a) a vulgar fraction in its lowest terms   b) a decimal.

4. Find 36% of £2.50.

5. There are 450 children in a primary school, 12% of whom do not speak English at home. Find the number of children for whom English is not their home language.

5 SCALE DRAWING

ACCURATE DRAWING WITH SCALED DOWN MEASUREMENTS

If you are asked to draw a car park which is a rectangle measuring 50 m by 25 m, you obviously cannot draw it full size. To fit it on to your page you will have to scale down the measurements. In this case you could use 1 cm to represent 5 m on the car park. This is called the *scale*; it is usually written as 1 cm ≡ 5 m, and must *always* be stated on any scale drawing.

EXERCISE 5a Start by making a rough drawing of the object you are asked to draw to scale. Mark all the full size measurements on your sketch. Next draw another sketch and put the scaled measurements on this one. Then do the accurate scale drawing.

> The end wall of a bungalow is a rectangle with a triangular top. The rectangle measures 6 m wide by 3 m high. The base of the triangle is 6 m and the sloping sides are 4 m long. Using a scale of 1 cm to 1 m, make a scale drawing of this wall. Use your drawing to find, to the nearest tenth of a metre, the distance from the ground to the ridge of the roof.

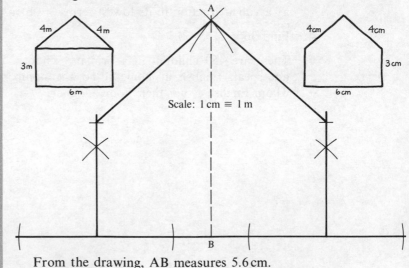

Scale: 1 cm ≡ 1 m

> From the drawing, AB measures 5.6 cm.
> So the height of the wall is 5.6 × 1 m = 5.6 m.

In questions 1 to 5, use a scale of 1 cm to 1 m.

1.

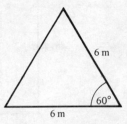

6 m

60°

6 m

4.

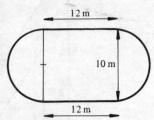

12 m

10 m

12 m

2.

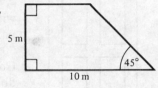

5 m

10 m

45°

5.

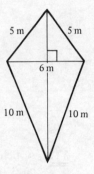

5 m   5 m

6 m

10 m   10 m

3.

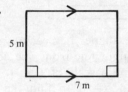

5 m

7 m

In questions 6 to 10, choose your own scale.

Choose a scale that gives lines that are long enough to draw easily; in general, the lines on your drawing should be at least 5 cm long. Avoid scales that give lengths involving awkward fractions of a centimetre, such as thirds; $\frac{1}{3}$ cm cannot be read from your ruler.

6.

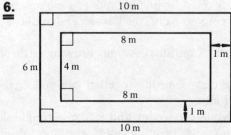

10 m

8 m

1 m

6 m   4 m

8 m

1 m

10 m

7.

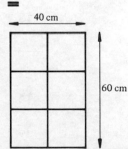

40 cm

60 cm

A casement window with equally spaced glazing bars

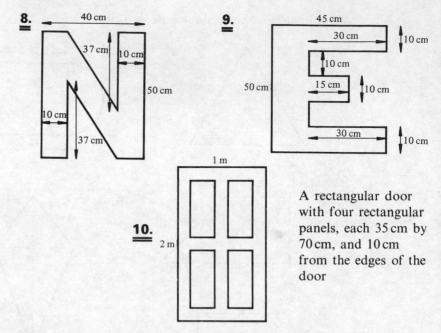

8.

9.

10.

A rectangular door with four rectangular panels, each 35 cm by 70 cm, and 10 cm from the edges of the door

11. A field is rectangular in shape. It measures 300 m by 400 m. A land drain goes in a straight line from one corner of the field to the opposite corner. Using a scale of 1 cm to 50 m, make a scale drawing of the field and use it to find the length of the land drain.

12. The end wall of a ridge tent is a triangle. The base is 2 m and the sloping edges are each 2.5 m. Using a scale of 1 cm to 0.5 m, make a scale drawing of the triangular end of the tent and use it to find the height of the tent.

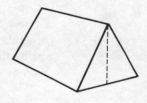

13. The surface of a swimming pool is a rectangle measuring 25 m by 10 m. Choose your own scale and make a scale drawing of the pool.
Now compare and discuss your drawing with other pupils.

14. The whole class working together can collect the information for this question.
Measure your classroom and make a rough sketch of the floor plan. Mark the position and width of doors and windows. Choosing a suitable scale, make an accurate scale drawing of the floor plan of your classroom.

ANGLES OF ELEVATION

If you are standing on level ground and can see a tall building, you will have to look up to see the top of that building.

If you start by looking straight ahead and then look up to the top of the building, the angle through which you raise your eyes is called the *angle of elevation* of the top of the building.

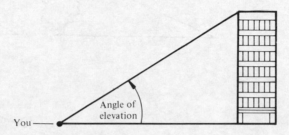

There are instruments for measuring angles of elevation. A simple one can be made from a large card protractor and a piece of string with a weight on the end.

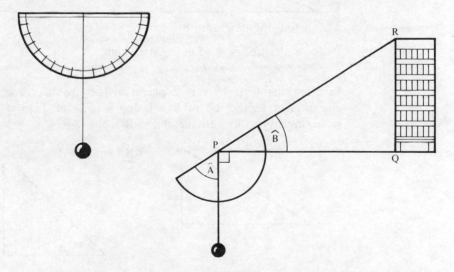

You can read the size of $\widehat{A}$.
Then the angle of elevation, $\widehat{B}$, is given by $\widehat{B} = 90° - \widehat{A}$.
(Note that this method is not very accurate.)

If your distance from the foot of the building and the angle of elevation of the top are both known, you can make a scale drawing of $\triangle PQR$.
Then this drawing can be used to work out the height of the building.

EXERCISE 5b

From a point, A, on the ground which is 50 m from the base of a tree, the angle of elevation of the top of the tree is 22°. Using a scale of 1 cm ≡ 5 m, make a scale drawing and use it to find the height of the tree.

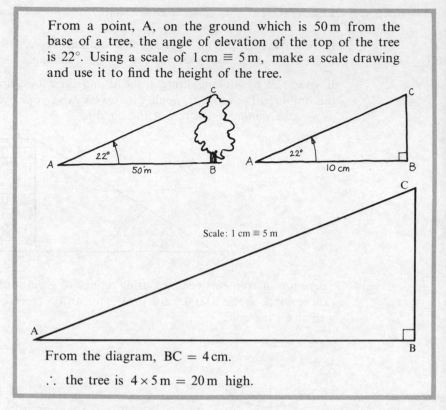

Scale: 1 cm ≡ 5 m

From the diagram, BC = 4 cm.

∴ the tree is 4 × 5 m = 20 m high.

In questions 1 to 4, A is a place on the ground, $\widehat{A}$ is the angle of elevation of C, the top of BC. Using a scale of 1 cm ≡ 5 m, make a scale drawing to find the height of BC.

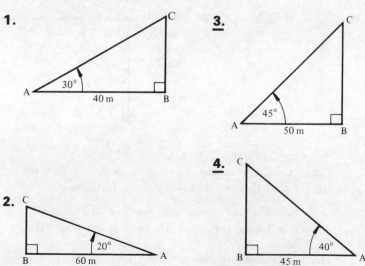

1.

A 30° 40 m B C

3.

A 45° 50 m B C

2.

C B 20° 60 m A

4.

C B 40° 45 m A

In questions 5 to 7, use a scale of 1 cm ≡ 10 m.

5. From A, the angle of elevation
of C is 35°. Find BC.

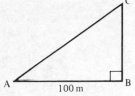

6. From P, the angle of elevation
of R is 15°. Find QR.

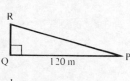

7. From N, the angle of elevation
of L is 30°. Find ML.

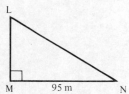

8. From a point, D, on the ground which is 100 m from the foot of
a church tower, the angle of elevation of the top of the tower is
30°. Use a scale of 1 cm to 10 m to make a scale drawing. Use
your drawing to find the height of the tower.

9. From the opposite side of the road, the angle of elevation of the
top of the roof of my house is 37°. The horizontal distance from
the point where I measured the angle to the middle of the house
is 12 m. Make a scale drawing, using a scale of 1 cm to 1 m, and
use it to find the height of the top of the roof.

10. From a point, P, on the ground which is 200 m from the base of
the Eiffel Tower, the angle of elevation of the top is 56°. Use a
scale of 1 cm to 20 m to make a scale diagram and find the
height of the Eiffel Tower.

11. From a point on the ground which is 300 m from the base of
the National Westminster Tower, the angle of elevation of the
top of the tower is 31°. Using a scale of 1 cm to 50 m, make a
scale drawing and find the height of the National Westminster
Tower. (This is a high office building in the City of London.)

12. The top of a radio mast is 76 m from the ground. From a point,
P, on the ground, the angle of elevation of the top of the mast
is 40°. Use a scale of 1 cm to 10 m to make a scale drawing to
find how far away P is from the mast.
(You will need to do some calculation before you can do the
scale drawing.)

ANGLES OF DEPRESSION

An *angle of depression* is the angle between the line looking straight ahead and the line looking *down* at an object below you.

If, for example, you are standing on a cliff looking out to sea, the diagram shows the angle of depression of a boat.

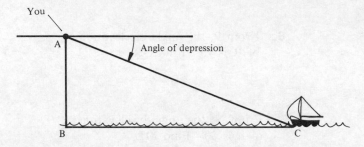

If the angle of depression and the height of the cliff are both known, you can make a scale drawing of △ABC. Then you can work out the distance of the boat from the foot of the cliff.

EXERCISE 5c

From the top of a cliff 20 m high, the angle of depression of a boat out at sea is 24°. Using a scale of 1 cm to 5 m, make a scale drawing to find the distance of the boat from the foot of the cliff.

Scale: 1 cm ≡ 5 m

From the diagram, BC = 9 cm.

∴ the distance of the boat from the foot of the cliff is

$$9 \times 5 \, \text{m} = 45 \, \text{m}$$

In questions 1 to 4, use a scale of 1 cm ≡ 10 m.

1.

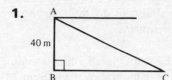

From A, the angle of depression
of C is 25°. Find BC.

2.

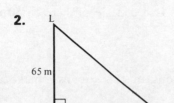

From L, the angle of depression
of N is 40°. Find MN.

3.

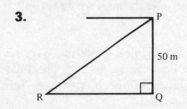

From P, the angle of depression
of R is 35°. Find RQ.

4.

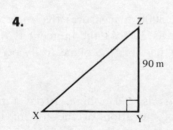

From Z, the angle of depression
of X is 42°. Find XY.

5. From the top of Blackpool Tower, which is 158 m high, the
angle of depression of a ship at sea is 25°. Use a scale of 1 cm
to 5 m to make a scale drawing to find the distance of the ship
from the base of the tower.

6. From the top of the Eiffel Tower, which is 300 m high, the angle
of depression of a house is 20°. Use a scale of 1 cm to 50 m to
make a scale drawing and find the distance of the house from
the base of the tower.

7. From the top of a vertical cliff, which is 30 m high, the angle of
depression of a yacht is 15°. Using a scale of 1 cm to 5 m, make
a scale drawing to find the distance of the yacht from the foot
of the cliff.

8. An aircraft flying at a height of 300 m measures the angle of depression of the end of runway as 18°. Using a scale of 1 cm to 100 m, make a scale diagram to find the horizontal distance of the aircraft from the runway.

9. The Sears Tower in Chicago is an office building and it is 443 m high. From the top of this tower, the angle of depression of a ship on a lake is 40°. How far away from the base of the building is the ship? Use a scale of 1 cm to 50 m to make your scale drawing.

For the remaining questions in this exercise, make a scale drawing choosing your own scale.

10. From a point on the ground 60 m away, the angle of elevation of the top of a factory chimney is 42°. Find the height of the chimney.

11. From the top of a hill, which is 400 m above sea level, the angle of depression of a boat house is 20°. The boat house is at sea level. Find the distance of the boat house from the top of the hill.

12. An aircraft flying at 5000 m measures the angle of depression of a point on the coast as 30°. At the moment that it measures the angle, how much further has the plane to fly before passing over the coast line?

13. A vertical radio mast is 250 m high. From a point, A, on the ground, the angle of elevation of the top of the mast is 30°. How far is the point A from the foot of the mast?

14. An automatic lightship is stationed 500 m from a point, A, on the coast. There are high cliffs at A and from the top of these cliffs, the angle of depression of the lightship is 15°. How high are the cliffs?

15. An airport controller measures the angle of elevation of an approaching aircraft as 20°. If the aircraft is then 1.6 km from the control building, at what height is it flying?

16. A surveyor, standing 400 m from the foot of a church tower, on level ground, measures the angle of elevation of the top of the tower. If this angle is 35° how high is the tower?

THREE FIGURE BEARINGS

A bearing is a compass direction.

If you are standing at a point, A, and looking at a tree, B, in the distance, as shown in the diagram below, then using compass directions you could say that

from A, the bearing of B is SE

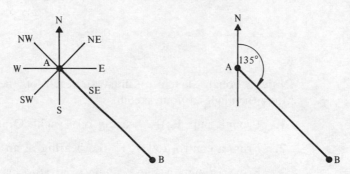

Using the modern method of a three figure bearing we first look north and then turn clockwise until we are looking at B. The angle turned through is the three figure bearing.
In this case

from A, the bearing of B is 135°

> A three figure bearing is a clockwise angle measured from the north.

If the angle is less than 100°, it is made into a three figure angle by putting zero in front, e.g. 20° becomes 020°.

EXERCISE 5d

Draw a rough sketch to illustrate that the bearing of a lighthouse, B, from a ship, A, is 060°. Mark the angle in your sketch.

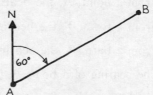

From a ship, C, the bearing of a ship, D, is 290°. Make a rough sketch and mark the angle.

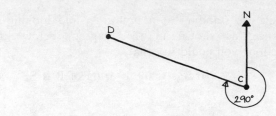

Draw a rough sketch to illustrate each of the following bearings. Mark the angle in your sketch.

1. From a ship, P, the bearing of a yacht, Q, is 045°

2. From a control tower, F, the bearing of an aeroplane, A, is 090°

3. From a point, A, the bearing of a radio mast, M, is 120°

4. From a town, T, the bearing of another town, S, is 180°

5. From a point, H, the bearing of a church, C, is 210°

6. From a ship, R, the bearing of a port, P, is 300°

7. From an aircraft, A, the bearing of an airport, L, is 320°

8. From a town, D, the bearing of another town, E, is 260°

9. From a helicopter, G, the bearing of a landing pad, P, is 060°

10. From a point, L, the bearing of a tree, T, is 270°

11. The bearing of a ship, A, from the pier, P, is 225°

12. The bearing of a radio mast, S, from a point, P, is 140°

13. The bearing of a yacht, Y, from a tanker, T, is 075°

14. The bearing of a town, Q, from a town, R, is 250°

15. The bearing of a tree, X, from a hill top, Y, is 025°

16. From a point, A, the bearing of a house, H, is 190°

17. From a town, T, the bearing of another town, S, is 290°

18. From a barn, B, the bearing of a tree, T, is 020°

19. The bearing of a boat, B, from a jetty, J, is 030°

20. The bearing of a flagpole, F, from a tent, T, is 300°

USING BEARINGS TO FIND DISTANCES

If we measure the bearing of a distant object from two different positions and then make a scale diagram, we can use this diagram to find the distance of that object from one or other of the positions.

EXERCISE 5e

From one end, A, of a road the bearing of a building, L, is 015°. The other end of the road, B, is 300 m due east of A. From B the bearing of the building is 320°. Using a scale of 1 cm to 50 m, make a scale diagram to find the distance of the building from A.

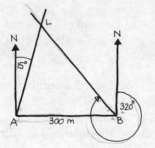

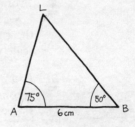

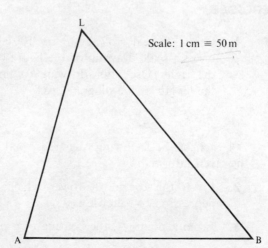

Scale: 1 cm ≡ 50 m

From the diagram, LA = 5.6 cm

∴ the distance of the building from A is 5.6 × 50 m = 280 m

1. From a point, A, the bearing of a tree, C, is 060°. From a second point B, which is 100 m due east of A, the bearing of the tree is 330°. Use a scale of 1 cm to 10 m to make a scale diagram and find the distance of the tree from A.

2. From a point, A, the bearing of a ship, C, is 140°. From a second point, B, which is 200 m due east of A, the bearing of the ship is 210°. Using a scale of 1 cm to 20 m to make a scale diagram and use it to find the distance of the ship from B.

3. From a point, A, the bearing of a tower, T, is 030°. From a second point, B, which is 400 m due north of A, the bearing of the tower is 140°. Using a scale of 1 cm to 50 m, make a scale drawing and use it to find the distance of the tower from A.

4. From a point, A, the bearing of a radar mast, M, is 060°. From a second point, B, which is 40 m due east of A, the bearing of the radar mast is 010°. Use a scale of 1 cm to 10 m and make a scale drawing to find the distance of the radar mast from A.

5. From a ship, P, the bearing of a submarine, S, is 020°. From a second ship, Q, which is 1000 m due north of P, the bearing of the submarine is 070°. Using a scale of 1 cm to 200 m, make a scale drawing to find the distance of the submarine from P.

MIXED EXERCISES

EXERCISE 5f   **1.** Using a scale of 1 cm to 100 cm, make a scale drawing of the figure on the right. Use your drawing to find the length of the diagonal AC.

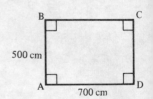

For each of the following questions, make a rough sketch to show all the given information.

2. From the top of a tower which is 150 m tall, the angle of depression of a house is 17°.

3. From a point, A, the bearing of a point, B is 270°.

4. An aircraft is flying at a height of 2000 m. From a point on the ground its angle of elevation is 40°.

5. An aircraft is flying at a height of 500 m when it measures the angle of depression of the end of the runway as 30°.

EXERCISE 5g   **1.**   Use a scale of 1 cm to 10 m to make a scale drawing of the figure on the right. Use your scale drawing to find the length of AC.

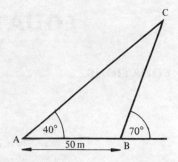

For each of the following questions, make a rough sketch showing all the given information.

2.   The bearing of a ship, R, from a ship, S, is 075°.

3.   From a position, A, on the ground, the angle of elevation of the top of an office block is 25°. The office block is 75 m tall.

4.   From the top of a cliff which is 50 m high, the angle of depression of a boat is 34°.

5.   From a ship, A, the bearing of an oil tanker, T, is 300°. From a second ship, B, which is 1000 m due west of A, the bearing of the oil tanker is 060°. Explain why the oil tanker is the same distance from A as it is from B.

EXERCISE 5h   **1.**   Use a scale of 1 cm to 50 m to make a scale drawing of the figure on the right. Use your drawing to find the distance CD.

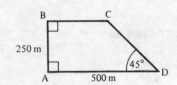

For the following questions, make rough sketches showing all the given information.

2.   From the top window of a house the angle of depression of the end of the garden is 32°. The garden is 10 m long.

3.   The bearing of an aircraft, X, from the control tower, T, is 157°.

4.   The angle of elevation of a helicopter, H, from the landing pad, L, is 45°. The helicopter is at a height of 45 m.

5.   From a position, A, the bearing of an army tank, T, is 210°. From a point, B, which is 50 m due south of A, the bearing of the tank is 280°. Which point is nearer to the tank, A or B?

6 EQUATIONS AND FORMULAE

EQUATIONS

Imagine a balance.

On the left-hand side there are two bags each containing the same (but unknown) number of apples and three loose apples.

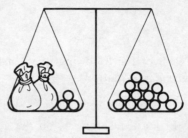

On the right-hand side there are thirteen apples.

Using the letter x to stand for the unknown number of apples in each bag we can write this as an equation:

$$2x + 3 = 13$$

We can solve this equation (i.e. find the number that x represents) as follows:

take three apples off each pan $\quad 2x = 10$

halve the contents of each pan $\quad x = 5$

EXERCISE 6a

Solve the equation $5x - 4 = 6$

$$5x - 4 = 6$$

Add 4 to each side $\qquad 5x = 10$

Divide each side by 5 $\qquad x = 2$

Check: LHS $= 5 \times 2 - 4 \quad$ RHS $= 6$
$\qquad\qquad = 6$

Solve the following equations:

1. $2x = 8$

2. $x - 3 = 1$

3. $x + 4 = 16$

4. $2x + 3 = 7$

5. $3x + 5 = 14$

6. $3x - 2 = 10$

7. $5 + 2x = 7$

8. $5x - 4 = 11$

9. $3 + 6x = 15$

10. $7x - 6 = 15$

Solve the equation $3x+4 = 12-x$

(We need to start by getting the terms containing x on one side of the equation and the terms without x on the other side. The left-hand side has the greater number of xs, so we will collect them on this side.)

$$3x+4 = 12-x$$

Add x to each side $\qquad\qquad 4x+4 = 12$

Take 4 from each side $\qquad\qquad 4x = 8$

Divide each side by 4 $\qquad\qquad x = \dfrac{8}{4} = 2$

Solve the equation $4-x = 6-3x$

(There are fewer x's missing from the LHS so we will collect them on this side.)

$$4-x = 6-3x$$

Add $3x$ to each side $\qquad\qquad 4-x+3x = 6$

$$4+2x = 6$$

Take 4 from each side $\qquad\qquad 2x = 2$

Divide each side by 2 $\qquad\qquad x = 1$

Solve the following equations:

11. $2x+5 = x+9$ **16.** $x+4 = 4x+1$

12. $3x+2 = 2x+7$ **17.** $3x-2 = 2x+1$

13. $x-4 = 2-x$ **18.** $1-3x = 9-4x$

14. $3-2x = 7-3x$ **19.** $2-5x = 6-3x$

15. $2x+1 = 4-x$ **20.** $5-3x = 1+x$

Solve the equation $4x+2-x = 7+x-3$

$$4x+2-x = 7+x-3$$

Collect like terms $\qquad\qquad 3x+2 = 4+x$

Take x from each side $\qquad\qquad 2x+2 = 4$

Take 2 from each side $\qquad\qquad 2x = 2$

Divide each side by 2 $\qquad\qquad x = 1$

Solve the following equations:

21. $x+2+2x = 8$

22. $x-4 = 3-x+1$

23. $3x+1-x = 5$

24. $4+3x-1 = 6$

25. $7+4x = 2-x+10$

26. $3+x-1 = 3x$

27. $x-4+2x = 5+x-1$

28. $x+5-2x = 3+x$

29. $x+17-4x = 2-x+6$

30. $8-3x-3 = x-4+2x$

31. $5x-8 = 2$

32. $4-x = 3x$

33. $5-x = 7+2x-4$

34. $4-2x = 8-4x$

35. $15 = 21-2x$

36. $x+4-3x = 2-x$

37. $3x-7 = 9-x+6$

38. $x+4 = 6x$

39. $8-3x = 5x$

40. $5-4x+7 = 2x$

BRACKETS

Reminder: If we want to multiply both x and 3 by 4, we group x and 3 together in a bracket and write $4(x+3)$.

So $4(x+3)$ means that *both x and* 3 are to be multiplied by 4. (Note that the multiplication sign is invisible, as it is in $5a$.)

i.e. $$4(x+3) = 4x+12$$

EXERCISE 6b Multiply out the following brackets:

1. $6(x+4)$

2. $3(2x+1)$

3. $4(x-3)$

4. $2(3x-5)$

5. $4(3-2x)$

6. $5(4x+2)$

7. $3(2-3x)$

8. $7(5-4x)$

9. $2(5x-7)$

10. $6(7+2x)$

Simplify $2(x-3)+4(3-2x)$

$2(x-3)+4(3-2x) = 2x-6+12-8x$ (brackets worked out)

$\qquad\qquad\qquad\quad = 6-6x$ (like terms collected)

Simplify:

11. $2(3+x)+3(2x+4)$

16. $3(3x+1)+4(x+4)$

12. $7(2x+3)+4(3x-2)$

17. $5(2x+3)+6(3x+2)$

13. $4(6x+3)+5(2x-5)$

18. $6(2x-5)+2(3x-7)$

14. $2(2x-4)+4(x+3)$

19. $8(2-x)+3(3+4x)$

15. $5(3x-2)+3(2x+5)$

20. $5(7-2x)+4(3-5x)$

21. $3(2x-1)+4(x+2)$

26. $5(4+3x)+3(2+7x)$

22. $5(2-x)+2(2x+1)$

27. $4(3+2x)+5(4-3x)$

23. $3(x-4)+7(2x-3)$

28. $8(x+1)+7(2-x)$

24. $2(2x+1)+4(3-2x)$

29. $3(2x+7)+5(3x-8)$

25. $6(2-x)+2(1-2x)$

30. $9(x-2)+5(4-3x)$

Solve the equation $3(4-x) = 9$

$$3(4-x) = 9$$

Multiply out the bracket $\qquad 12-3x = 9$

Add $3x$ to each side $\qquad 12 = 9+3x$

Take 9 from each side $\qquad 3 = 3x$

Divide each side by 3 $\qquad 1 = x \quad$ i.e. $\quad x = 1$

Solve the following equations:

31. $2(x+2) = 8$

35. $3(2x+5) = 18$

32. $4(2-x) = 2$

36. $3(3-2x) = 3$

33. $5(3x+1) = 20$

37. $2(x+4) = 3(2x+1)$

34. $2(2x-1) = 6$

38. $4(2x-3) = 2(3x-5)$

39. $6(3x+5) = 12$

43. $5(3-2x) = 3(4-3x)$

40. $6(x+3) = 2(2x+5)$

44. $7(1+2x) = 21$

41. $8(x-1) = 4$

45. $7(2x-1) = 5(3x-2)$

42. $3(1-4x) = 11$

46. $4(3x+2) = 14$

MULTIPLICATION AND DIVISION OF FRACTIONS

Remember that, to multiply fractions, the numerators are multiplied together and the denominators are multiplied together:

i.e.
$$\frac{3}{4} \times \frac{5}{7} = \frac{3 \times 5}{4 \times 7} = \frac{15}{28}$$

Also $\frac{1}{6}$ of x means $\frac{1}{6} \times x = \frac{1}{6} \times \frac{x}{1} = \frac{x}{6}$ (1)

Remember that, to divide by a fraction, that fraction is turned upside down and multiplied:

i.e.
$$\frac{2}{3} \div \frac{5}{7} = \frac{2}{3} \times \frac{7}{5} = \frac{14}{15}$$

and
$$x \div 6 = \frac{x}{1} \div \frac{6}{1} = \frac{x}{1} \times \frac{1}{6} = \frac{x}{6}$$ (2)

Comparing (1) and (2) we see that

$\frac{1}{6}$ of x, $\frac{1}{6}x$, $x \div 6$ and $\dfrac{x}{6}$ are all equivalent

EXERCISE 6c

Simplify $12 \times \dfrac{x}{3}$

$$12 \times \frac{x}{3} = \frac{\overset{4}{\cancel{12}}}{1} \times \frac{x}{\cancel{3}_1}$$

$$= 4x$$

Simplify $\dfrac{2x}{3} \div 8$

$$\frac{2x}{3} \div 8 = \frac{2x}{3} \div \frac{8}{1}$$

$$= \frac{\overset{1}{\cancel{2x}}}{3} \times \frac{1}{\cancel{8}_4}$$ (Remember that $2x = 2 \times x$)

$$= \frac{x}{12}$$

Simplify:

1. $4 \times \dfrac{x}{8}$ **6.** $\dfrac{1}{5}$ of $10x$

2. $\dfrac{1}{2} \times \dfrac{x}{3}$ **7.** $\dfrac{2}{5} \times \dfrac{3x}{4}$

3. $9 \times \dfrac{x}{6}$ **8.** $\dfrac{3}{4} \times 2x$

4. $\dfrac{1}{3}$ of $2x$ **9.** $\dfrac{2}{3}$ of $9x$

5. $\dfrac{2x}{3} \times \dfrac{6}{5}$ **10.** $\dfrac{x}{2} \times \dfrac{x}{3}$

11. $\dfrac{5x}{2} \div 4$ **16.** $\dfrac{3}{4} \times \dfrac{2x}{5}$

12. $\dfrac{4x}{9} \div 8$ **17.** $\dfrac{4x}{9} \div \dfrac{2}{3}$

13. $\dfrac{x}{3} \div \dfrac{1}{6}$ **18.** $\dfrac{3}{5}$ of $15x$

14. $\dfrac{x}{4} \div \dfrac{1}{2}$ **19.** $\dfrac{3x}{2} \div \dfrac{1}{6}$

15. $\dfrac{2x}{3} \div \dfrac{5}{6}$ **20.** $\dfrac{5x}{3} \times \dfrac{6x}{15}$

FRACTIONAL EQUATIONS

EXERCISE 6d

Solve the equation $\dfrac{x}{3} = 2$

(As $\dfrac{x}{3}$ means $\dfrac{1}{3}$ of x, to find x we need to make $\dfrac{x}{3}$ three times larger.)

$$\dfrac{x}{3} = 2$$

Multiply each side by 3 $\dfrac{x}{\cancel{3}} \times \dfrac{\cancel{3}^{1}}{1} = 2 \times 3$

$$x = 6$$

Solve the following equations:

1. $\dfrac{x}{5} = 3$　　　　　　　　　**6.** $\dfrac{2x}{5} = 9$

2. $\dfrac{x}{2} = 4$　　　　　　　　　**7.** $\dfrac{4x}{7} = 8$

3. $\dfrac{x}{6} = 8$　　　　　　　　　**8.** $\dfrac{6x}{5} = 10$

4. $\dfrac{2x}{3} = 8$　　　　　　　　　**9.** $12 = \dfrac{3x}{2}$

5. $16 = \dfrac{9x}{2}$　　　　　　　　**10.** $\dfrac{3x}{4} = 6$

Solve the equation $\dfrac{2x}{5} = \dfrac{1}{3}$

$$\frac{2x}{5} = \frac{1}{3}$$

Multiply each side by 5

$$\frac{2x}{\cancel{5}_1} \times \frac{\cancel{5}^1}{1} = \frac{1}{3} \times \frac{5}{1}$$

$$2x = \frac{5}{3}$$

Divide each side by 2

$$x = \frac{5}{3} \div 2$$

$$x = \frac{5}{3} \times \frac{1}{2}$$

$$x = \frac{5}{6}$$

Solve the following equations:

11. $\dfrac{3x}{2} = \dfrac{1}{4}$　　　　　　　**14.** $\dfrac{6x}{5} = \dfrac{2}{3}$

12. $\dfrac{4x}{3} = \dfrac{1}{5}$　　　　　　　**15.** $\dfrac{3x}{8} = \dfrac{1}{2}$

13. $\dfrac{2x}{9} = \dfrac{1}{3}$　　　　　　　**16.** $\dfrac{5x}{7} = \dfrac{3}{4}$

17. $\dfrac{3x}{5} = \dfrac{1}{4}$

19. $\dfrac{2x}{9} = \dfrac{4}{5}$

18. $\dfrac{4x}{7} = \dfrac{2}{5}$

20. $\dfrac{6x}{11} = \dfrac{5}{7}$

Solve the equation $\dfrac{x}{5} + \dfrac{1}{2} = 1$

(Both 5 and 2 divide into 10, so by multiplying each side by 10 we can eliminate all fractions from this equation before we start to solve for x.)

$$\frac{x}{5} + \frac{1}{2} = 1$$

Multiply both sides by 10

$$10\left(\frac{x}{5} + \frac{1}{2}\right) = 10 \times 1$$

$$\frac{\overset{2}{\cancel{10}}}{1} \times \frac{x}{\cancel{5}_1} + \frac{\overset{5}{\cancel{10}}}{1} \times \frac{1}{\cancel{2}_1} = 10$$

$$2x + 5 = 10$$

Take 5 from each side

$$2x = 5$$

Divide each side by 2

$$x = 2\tfrac{1}{2}$$

Solve the following equations:

21. $\dfrac{x}{3} + \dfrac{1}{4} = 1$

26. $\dfrac{x}{3} + \dfrac{5}{6} = 2$

22. $\dfrac{x}{5} - \dfrac{3}{4} = 2$

27. $\dfrac{x}{3} - \dfrac{2}{9} = 4$

23. $\dfrac{x}{5} + \dfrac{2x}{3} = 3$

28. $\dfrac{3x}{4} - \dfrac{x}{2} = 5$

24. $\dfrac{5x}{7} + \dfrac{x}{2} = 2$

29. $\dfrac{3}{4} - \dfrac{x}{5} = 1$

25. $\dfrac{2x}{3} - \dfrac{1}{2} = 4$

30. $\dfrac{5}{7} + \dfrac{3x}{4} = 2$

Solve the equation $\dfrac{x}{2} = \dfrac{1}{6} + \dfrac{x}{3}$

(2, 3 and 6 all divide into 6, so multiplying each side by 6 will eliminate all fractions from this equation.)

$$\frac{x}{2} = \frac{1}{6} + \frac{x}{3}$$

Multiply each side by 6

$$\frac{x}{2} \times \frac{6}{1} = \left(\frac{1}{6} + \frac{x}{3}\right) \times \frac{6}{1}$$

$$\frac{x}{2} \times \frac{6}{1} = \frac{1}{6} \times \frac{6}{1} + \frac{x}{3} \times \frac{6}{1}$$

$$3x = 1 + 2x$$

Take $2x$ from each side $x = 1$

Check: LHS $= \dfrac{1}{2}$ RHS $= \dfrac{1}{6} + \dfrac{1}{3}$

$$= \frac{1}{6} + \frac{2}{6} = \frac{3}{6} = \frac{1}{2}$$

Solve the following equations:

31. $\dfrac{x}{3} + \dfrac{1}{4} = \dfrac{1}{2}$

32. $\dfrac{x}{5} + \dfrac{2}{3} = \dfrac{14}{15}$

33. $\dfrac{x}{4} - \dfrac{1}{2} = \dfrac{9}{4}$

34. $\dfrac{2x}{3} + \dfrac{2}{7} = \dfrac{1}{3}$

35. $\dfrac{x}{2} - \dfrac{3}{7} = \dfrac{1}{2}$

36. $\dfrac{3x}{5} + \dfrac{2}{9} = \dfrac{11}{15}$

37. $\dfrac{5x}{6} + \dfrac{x}{8} = \dfrac{3}{4}$

38. $\dfrac{3x}{4} + \dfrac{1}{8} = \dfrac{1}{2}$

39. $\dfrac{5x}{12} - \dfrac{1}{3} = \dfrac{x}{8}$

40. $\dfrac{2x}{5} - \dfrac{x}{15} = \dfrac{5}{9}$

41. $\dfrac{3x}{4} + \dfrac{1}{3} = \dfrac{x}{2} + \dfrac{5}{8}$

42. $\dfrac{2x}{7} - \dfrac{3}{4} = \dfrac{x}{14} + \dfrac{1}{2}$

43. $\dfrac{5x}{7} - \dfrac{2}{3} = \dfrac{3}{7} - \dfrac{x}{3}$

44. $\dfrac{2x}{9} - \dfrac{3}{4} = \dfrac{7}{18} - \dfrac{5x}{12}$

45. $\dfrac{3}{11} - \dfrac{x}{2} = \dfrac{2x}{11} + \dfrac{1}{4}$ **48.** $\dfrac{x}{3} + \dfrac{1}{4} - \dfrac{x}{6} = \dfrac{7}{12}$

46. $\dfrac{3}{5} - \dfrac{x}{9} = \dfrac{2}{15} - \dfrac{2x}{45}$ **49.** $\dfrac{5}{8} - \dfrac{x}{6} + \dfrac{1}{12} = \dfrac{3}{4}$

47. $\dfrac{4}{7} + \dfrac{2x}{9} = \dfrac{15}{9} - \dfrac{4x}{21}$ **50.** $\dfrac{5}{9} - \dfrac{7x}{12} = \dfrac{1}{6} - \dfrac{x}{8}$

PROBLEMS

EXERCISE 6e   Form an equation for each of the following problems and then solve the equation.

> A bag of sweets was divided into three equal shares. David had one share and he got 8 sweets. How many sweets were there in the bag?
>
> Let x stand for the number of sweets in the bag.
>
> One share is $\frac{1}{3}$ of x $\therefore$ $\frac{1}{3}$ of $x = 8$
>
> $$\dfrac{x}{3} = 8$$
>
> Multiply each side by 3 $x = 24$
>
> Therefore there were 24 sweets in the bag.

1. Tracy Brown came first in the Newtown Golf Tournament and won £100. This was $\frac{2}{3}$ of the total prize money paid out. Find the total prize money.

2. Peter lost 8 marbles in a game. This number was one-fifth of the number that he started with. Find how many he started with.

3. The width of a rectangle is 12 cm. This is two-fifths of its length. Find the length of the rectangle.

4. I think of a number, halve it and the result is 6. Find the number that I first thought of.

5. The length of a rectangle is 8 cm and this is $\frac{1}{3}$ of its perimeter. Find its perimeter.

6. In an equilateral triangle, the perimeter is 15 cm. Find the length of one side of the triangle.

7. I think of a number, take $\frac{1}{3}$ of it and then add 4. The result is 7. Find the number I first thought of.

8. I think of a number and divide it by 3. The result is 2 less than the number I first thought of. Find the number I first thought of.

9. I think of a number and add $\frac{1}{3}$ of it to $\frac{1}{2}$ of it. The result is 10. Find the number I first thought of.

10. John Smith won the singles competition of a local tennis tournament, for which he got $\frac{1}{5}$ of the total prize money. He also won the doubles competition for which he got $\frac{1}{20}$ of the prize money. He got £250 altogether. How much was the total prize money?

DIRECTED NUMBERS

Reminder:
$$(+2) \times (+3) = +6$$
$$(+2) \times (-3) = -6$$
$$(-2) \times (+3) = -6$$
$$(-2) \times (-3) = +6$$

EXERCISE 6f   Evaluate:

1. $(+2) \times (-4)$ **6.** $(-4) \times (-7)$

2. $(-3) \times (-5)$ **7.** $(-\frac{1}{3}) \times (-6)$

3. $(-6) \times (+4)$ **8.** $(+\frac{1}{2}) \times (+\frac{2}{3})$

4. $(-\frac{1}{2}) \times (+6)$ **9.** $(-\frac{3}{4}) \times (+12)$

5. $(+\frac{3}{4}) \times (+16)$ **10.** $(+5) \times (-9)$

Remember that the positive sign is often omitted, i.e. 6 means +6.

Simplify $4(x-3) - 3(2-3x)$

Multiply out the brackets

$$4(x-3) - 3(2-3x) = 4x - 12 - 6 + 9x$$

Collect like terms
$$= 13x - 18$$

Simplify:

11. $7-2(x-5)$

12. $2x+5(3x-4)$

13. $3x-6(3x+5)$

14. $4-7(2x-3)$

15. $3x-4(5-3x)$

16. $3(x-4)+6(3-2x)$

17. $2(3x+5)-2(4+3x)$

18. $5(2x-8)-3(2-5x)$

19. $7(x-2)-(2x+3)$

20. $5(4x-5)-(4-2x)$

Solve the equation $x-3(x-2)=8$

$$x-3(x-2)=8$$

Multiply out the brackets $x-3x+6=8$

Collect like terms $-2x+6=8$

Add $2x$ to each side $6=8+2x$

Take 8 from each side $-2=2x$

Divide each side by 2 $-1=x$ i.e. $x=-1$

Solve the following equations:

21. $4x-2(x-3)=8$

22. $7-3(5-2x)=10$

23. $4x+2(2x-5)=6$

24. $3(x-4)-7=2(x-3)$

25. $4-3x=3+4(2x-3)$

26. $3x-2(4-5x)=5-3x$

27. $2x-\frac{1}{2}(6+2x)=7$

28. $10-\frac{1}{4}(4x-8)=5$

29. $3-\frac{2}{3}(6x+9)=5-2x$

30. $\frac{3}{4}(4-8x)=2x-\frac{2}{3}(6-12x)$

FORMULAE

For all rectangles it is true that the area is equal to the length multiplied by the breadth, provided that the length and breadth are measured in the same unit.

If we use letters for the unknown quantities (A for area, l for length, b for breadth) we can write the first sentence more briefly as a formula: $A = l \times b$.

The multiplication sign is usually left out giving

$$A = lb$$

EXERCISE 6g   The letters in the diagrams all stand for a number of centimetres.

The perimeter of the square below is P cm. Write down a formula for P.

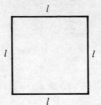

$$P = l+l+l+l$$

Collect like terms $P = 4l$

In each of the following figures the perimeter is P cm. Write down a formula for P starting with $P =$

1.

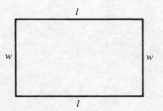

2.

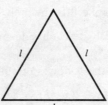

3.

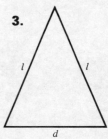

4.

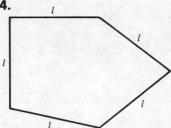

5.

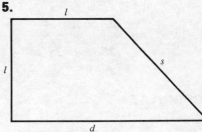

If G is the number of girls in a class and B is the number of boys, write down a formula for the total number, T, of children in the class.

$$T = G+B$$

6. I buy x lb of apples and y lb of pears. Write down a formula for W if W lb is the weight of fruit that I have bought.

7. If l m is the length of a rectangle and b m is the breadth, write down a formula for P if the perimeter of the rectangle is P m.

8. I start a game with N marbles and win another M marbles. Write down a formula for the number, T, of marbles that I finish the game with.

9. I start a game with N marbles and lose L marbles. Write down a formula for the number, T, of marbles that I finish with.

10. The side of a square is l m long. Write down a formula for A if the area of the square is A m^2.

11. Peaches cost n p each. Write down a formula for N if the cost of 10 peaches is N p.

12. Oranges cost x p each and I buy n of these oranges. Write down a formula for C where C p is the total cost of the oranges.

13. I have a piece of string which is l cm long. I cut off a piece which is d cm long. Write down a formula for L if the length of string which is left is L cm.

14. A rectangle is $2l$ m long and l m wide. Write down a formula for P where P m is the perimeter of the rectangle.

15. Write down a formula for A where A m^2 is the area of the rectangle described in question 14.

16. I had a bag of sweets with S sweets in it; I then ate T of them. Write down a formula for the number, N, of sweets left in the bag.

17. A lorry weighs T tonnes when empty. Steel girders weighing a total of S tonnes are then loaded on to the lorry. Write down a formula for W where W tonnes is the weight of the loaded lorry.

18. I started the term with a new packet of N felt tipped pens. During the term I lost L of them and R of them ran dry. Write down a formula for the number, S, that I had at the end of the term.

19. A train travels p km in one direction and then it comes back q km in the opposite direction. If it is then r km from its starting point, write down a formula for r.

20. One box of tinned fruit weights K kg. The weight of n such boxes is W kg. Write down a formula for W.

21. Two points have the same y-coordinate. The x-coordinate of one point is a and the x-coordinate of the other point is b. If d is the distance between the two points, write down a formula for d given that a is less than b. Make a sketch to illustrate this problem.

22. A letter costs x pence to post. The cost of posting 20 such letters is £q. Write down a formula for q. (Be careful—look at the units given.)

23. One grapefruit costs y pence. The cost of n such grapefruit is £L. Write down a formula for L. (Look carefully at the units.)

24. A rectangle is l m long and b cm wide. The area is A cm^2. Write down a formula for A.

25. On my way to work this morning the train I was travelling on broke down. I spent t hours on the train and s minutes walking. Write down a formula for T if the total time that my journey took was T hours.

SUBSTITUTING NUMERICAL VALUES INTO A FORMULA ──────

The formula for the area of a rectangle is $A = lb$.

If a rectangle is 3 cm long and 2 cm wide, we can substitute the number 3 for l and the number 2 for b to give $A = 3 \times 2 = 6$.

So the area of that rectangle is 6 cm^2.

When you substitute numerical values into a formula you may have a mixture of operations, i.e. ( ), $\times$, $\div$, $+$, $-$, to perform. Remember the order from the capital letters of "Bless My Dear Aunt Sally".

EXERCISE 6h

If $v = u + at$, find v when $u = 2$, $a = \frac{1}{2}$ and $t = 4$.

$$v = u + at$$

When $u = 2$, $a = \frac{1}{2}$, $t = 4$,

$$v = 2 + \frac{1}{2} \times 4$$
$$= 2 + 2$$
$$= 4$$

1. If $N = T + G$, find N when $T = 4$ and $G = 6$.

2. If $T = np$, find T when $n = 20$ and $p = 5$.

3. If $P = 2(l+b)$, find P when $l = 6$ and $b = 9$.

4. If $L = x - y$, find L when $x = 8$ and $y = 6$.

5. If $N = 4(l-s)$, find N when $l = 7$ and $s = 2$.

6. If $S = n(a+b)$, find S when $n = 20$, $a = 2$ and $b = 8$.

7. If $V = lbw$, find V when $l = 4$, $b = 3$ and $w = 2$.

8. If $A = \dfrac{PRT}{100}$, find A when $P = 100$, $R = 3$ and $T = 5$.

9. If $w = u(v - t)$, find w when $u = 5$, $v = 7$ and $t = 2$.

10. If $s = \frac{1}{2}(a+b+c)$, find s when $a = 5$, $b = 7$ and $c = 3$.

If $v = u - at$, find v when $u = 5$, $a = -2$, $t = -3$

$$v = u - at$$

When $u = 5$, $a = -2$, $t = -3$,

$$
\begin{aligned}
v &= 5 - (-2) \times (-3) \\
&= 5 - (+6) \\
&= 5 - 6 \\
&= -1
\end{aligned}
$$

(Notice that where negative numbers are substituted for letters they have been put in brackets. This makes sure that only one operation at a time is carried out.)

---

11. If $N = p + q$, find N when $p = 4$ and $q = -5$.

12. If $C = RT$, find C when $R = 4$ and $T = -3$.

13. If $z = w + x - y$, find z when $w = 4$, $x = -3$ and $y = -4$.

14. If $r = u(v - w)$, find r when $u = -3$, $v = -6$ and $w = 5$.

15. Given that $X = 5(T - R)$, find X when $T = 4$ and $R = -6$.

16. Given that $P = d - rt$, find P when $d = 3$, $r = -8$ and $t = 2$.

17. Given that $v = l(a+n)$, find v when $l = -8$, $a = 4$ and $n = -6$.

18. If $D = \dfrac{a-b}{c}$, find D when $a = -4$, $b = -8$ and $c = 2$.

19. If $Q = abc$, find Q when $a = 3$, $b = -7$ and $c = -5$.

20. If $I = \frac{2}{3}(x+y-z)$, find I when $x = 4$, $y = -5$ and $z = -6$.

Given that $2S = d(a+l)$, find a when $S = 20$, $d = 2$ and $l = 16$

$$2S = d(a+l)$$

Substituting $s = 20$, $d = 2$, $l = 16$ gives

$$40 = 2(a+16)$$

(We can now solve this equation for a.)

Multiply out the brackets $40 = 2a + 32$

Take 32 from each side $8 = 2a$

Divide by 2 $4 = a$ or $a = 4$

---

21. Given that $N = G + B$, find B when $N = 40$ and $G = 25$.

22. If $R = t \div c$, find t when $R = 10$ and $c = 20$.

23. Given that $d = st$, find t when $d = 50$ and $s = 15$.

24. If $N = 2(p+q)$, find q when $N = 24$, and $p = 5$.

25. Given that $L = P(2-a)$, find a when $L = 10$ and $P = 40$.

26. Given that $s = \frac{1}{3}(a-b)$, find b when $s = 15$ and $a = 24$.

27. Given that $v = u + at$, find u when $v = 32$, $a = 8$ and $t = 4$.

28. If $v^2 = u^2 + 2as$, find a when $v = 3$, $u = 2$ and $s = 12$.

29. If $d = \frac{1}{2}(a+b+c)$, find a when $d = 16$, $b = 4$ and $c = -3$.

30. If $H = P(Q-R)$, find Q when $H = 12$, $P = 4$ and $R = -6$.

PROBLEMS

1. Given that $v = at$, find the value of
a) v when $a = 4$ and $t = 12$
b) v when $a = -3$ and $t = 6$
c) t when $v = 18$ and $a = 3$
d) a when $v = 25$ and $t = 5$

2. Given that $N = 2(n-m)$, find the value of
a) N when $n = 6$ and $m = 4$
b) N when $n = 7$ and $m = -3$
c) n when $N = 12$ and $m = 2$
d) m when $N = 16$ and $n = -4$

3. If $A = P+QT$, find the value of
a) A when $P = 50$, $Q = \frac{1}{2}$ and $T = 4$
b) A when $P = 70$, $Q = 5$ and $T = -10$
c) P when $A = 100$, $Q = \frac{1}{4}$ and $T = 16$
d) T when $A = 25$, $P = -15$ and $Q = -10$

4. Given that $s = \frac{1}{2}(a-b)$, find the value of
a) s when $a = 16$ and $b = 6$
b) s when $a = -4$ and $b = -10$
c) a when $s = 15$ and $b = 8$
d) b when $s = 10$ and $a = -4$

5. Given that $z = x-3y$, find the value of
a) z when $x = 3\frac{1}{2}$ and $y = \frac{3}{4}$
b) z when $x = \frac{3}{8}$ and $y = -1\frac{1}{2}$
c) x when $z = 5\frac{1}{3}$ and $y = 2\frac{1}{2}$
d) y when $z = \frac{1}{4}$ and $x = \frac{7}{8}$

6. If $P = 100r-t$, find the value of
a) P when $r = 0.25$ and $t = 10$
b) P when $r = 0.145$ and $t = 15.6$
c) t when $P = 18.5$ and $r = 0.026$
d) r when $P = 50$ and $t = -12$

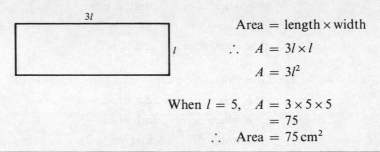

A rectangle is $3l$ cm long and l cm wide. If the area of the rectangle is A cm², write down a formula for A.
Use your formula to find the area of this rectangle if it is 5 cm wide.

Area = length × width

$\therefore \quad A = 3l \times l$

$A = 3l^2$

When $l = 5$, $\quad A = 3 \times 5 \times 5$
$= 75$
$\therefore \quad$ Area $= 75$ cm²

7. Oranges cost n p each. If the cost of a box of 50 of these oranges is C p, write down a formula for C. Use your formula to find the cost of a box of oranges if each orange costs 12 p.

8. Lemons cost n p each. The cost of a box of 50 lemons is £L. Write down a formula for L (be careful with the units). Use your formula to find the cost of a box of these lemons when they cost 10 p each.

9. A rectangular box is l cm long, b cm wide and d cm deep. The volume of the box is V cm³. Write down a formula for V. Use your formula to find the volume of a box measuring 20 cm by 12 cm by 5 cm.

10. A rectangle is a cm long and b cm wide. Write down a formula for P if P cm is the perimeter of the rectangle. Use your formula to find the perimeter of a rectangle measuring 20 cm by 15 cm.

11. The length of a rectangle is twice its width. If the rectangle is x cm wide, write down a formula for P if its perimeter is P cm. Use your formula to find the width of a rectangle that has a perimeter of 36 cm.

12. A roll of paper is L m long. N pieces each of length r m are cut off the roll. If the length of paper left is P m, write down a formula for P. Use your formula to find the length of paper left from a roll that was 20 m long after 10 pieces, each of length 1.5 m, are cut off.

13. An equilateral triangle has sides each of length a cm. I. perimeter of the triangle is P cm, write down a formula for P. Use your formula to find the lengths of the sides of an equilateral triangle whose perimeter is 72 cm.

14. Tins of baked beans weigh a g each. N of these tins are packed into a box. The empty box weighs p g. Write down a formula for W where W g is the weight of the full box. Use your formula to find the number of tins that are in a full box if the full box weighs 10 kg, the empty box weighs 1 kg and each tin weighs 200 g.

15. The rectangular box in the diagram is l cm long, w cm wide and h cm high. Write down a formula for A if A cm^2 is the total surface area of the box (i.e. the area of all six faces). Use your formula to find the surface area of a rectangular box measuring 50 cm by 30 cm by 20 cm.

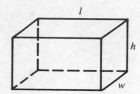

CHANGING THE SUBJECT OF A FORMULA

Suppose that we have to use the formula $A = lb$ to find the value of l when $A = 20$ and $b = 5$. There are two ways of doing this. Either we can substitute the numbers directly, giving $20 = l \times 5$ and solve this equation for l, which gives $l = 4$

Or, by dividing both sides of the formula by b, we can rearrange the formula to $l = \dfrac{A}{b}$,

then substitute in the numbers to give $\qquad l = \dfrac{20}{5} = 4$

When the formula is in the form $A = lb$, A is called the subject of the formula.

When the formula is in the form $l = \dfrac{A}{b}$, l is called the subject of the formula.

Changing from $A = lb$ to $l = \dfrac{A}{b}$ is called changing the subject of the formula.

EXERCISE 6j

Make r the subject of the formula $p = q+r$

(To make r the subject of $p = q+r$ we have to "solve" the formula for r.)

$$p = q+r$$

Take q from both sides $p-q = r$

or $r = p-q$

Make the letter in brackets the subject of the following formulae:

1. $N = T+G$ (T) **6.** $v = u+t$ (u)

2. $z = xy$ (x) **7.** $S = d-t$ (d)

3. $S = \dfrac{d}{t}$ (d) **8.** $P = 2y+z$ (z)

4. $L = X-Y$ (X) **9.** $C = RT$ (T)

5. $s = a+2b$ (a) **10.** $L = a+b+c$ (a)

11. $P = a+b$ (a) **16.** $x = y-z$ (y)

12. $N = R+T$ (T) **17.** $P = ab+c$ (c)

13. $b = a+c+d$ (c) **18.** $L = \dfrac{m}{n}$ (m)

14. $v = rt+u$ (u) **19.** $v = u+at$ (u)

15. $N = rn$ (n) **20.** $s = ax+y$ (y)

MIXED EXERCISES

EXERCISE 6k   **1.** Solve the equation $8 = 3+2x$.

2. Solve the equation $x-4 = 5-2x+1$.

3. Multiply out $3(2x-8)$.

4. Find $\dfrac{3}{5}$ of $10x$.

5. Solve the equation $\dfrac{2x}{3} = 8$.

6. Find the value of x if $\dfrac{x}{2}+\dfrac{1}{6}=\dfrac{1}{3}$.

7. Simplify $3x-2(4-x)$.

8. Write down a formula for P if P cm is the perimeter of the figure in the diagram. (Each letter stands for a number of centimetres.)

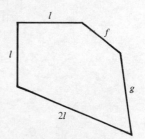

9. If $P = a-b$, find the value of P when $a = 2$ and $b = 5$.

10. Make N the subject of the formula $R = N-D$.

EXERCISE 6I

1. Solve the equation $3-x = 2+2x$.

2. Solve the equation $3(2x+2) = 10$.

3. Simplify $\frac{3}{4}\times 8x$.

4. Simplify $5x\div\frac{2}{3}$.

5. Solve the equation $\dfrac{3x}{5}=\dfrac{9}{10}$.

6. Simplify $6(3-2x)-4(2-x)$.

7. Solve the equation $\dfrac{x}{4}-\dfrac{3}{5}=\dfrac{7}{8}$.

8. If $z = x-2y$, find z when $x = 3$ and $y = -6$.

9. There are three classes in the first year of Appletown School. There are a children in one class, b children in another class and c children in the third class. Write down a formula for the number, N, of children in the first year.

10. Make N, the subject of the formula $n = N-ab$.

EXERCISE 6m **1.** Find $\frac{3}{8}$ of $10x$.

2. Solve the equation $5(3-4x) = x-2(3x-5)$.

3. I think of a number and double it, then I add on 3 and double the result: this gives 14. If x stands for the number I first thought of, form an equation for x and then solve it.

4. Simplify $\dfrac{3x}{4} \div \dfrac{9}{11}$.

5. Find $\frac{3}{8}$ of $\frac{2}{3}x$.

6. Simplify $5x - \frac{2}{3}(6x-9)$.

7. Solve the equation $\dfrac{3}{8} - \dfrac{5x}{6} = \dfrac{2}{3}$.

8. Given that $r = s-vt$, find the value of r when $s = 4$, $v = 3$ and $t = -2$.

9. A rectangle is twice as long as it is wide. If it is a cm wide, write down a formula for P where P cm is the perimeter of the rectangle.

10. Make p the subject of the formula $L = 3pq$.

7 COORDINATES AND THE STRAIGHT LINE

THE EQUATION OF A STRAIGHT LINE

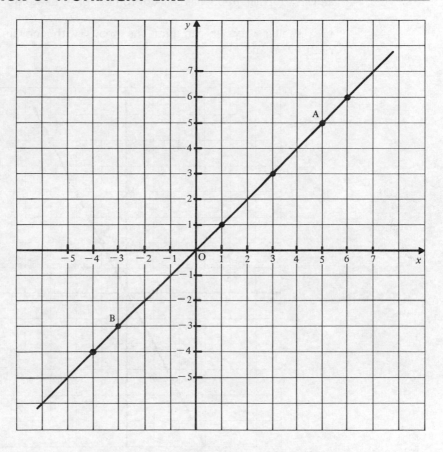

If we plot the points with coordinates $(-4, -4)$, $(1, 1)$, $(3, 3)$ and $(6, 6)$, we can see that a straight line can be drawn through these points which also passes through the origin.

For each point the y-coordinate is the same as the x-coordinate.

This is also true for any other point on this line,

e.g. the coordinates of A are $(5, 5)$ and of B are $(-3, -3)$.

Hence $\qquad y$-coordinate $= x$-coordinate

or simply $\qquad\qquad y = x$

This is called the equation of the line.

95

We can also think of a line as a set of points, i.e. this line is the set of points, or *ordered number pairs*, such that $\{(x, y)\}$ satisfies the relation $y = x$.

It follows that if another point on the line has an x-coordinate of -5, then its y-coordinate is -5
and if a further point has a y-coordinate of 4, its x-coordinate is 4.

In a similar way we can plot the points with coordinates $(-2, -4)$, $(1, 2)$, $(2, 4)$ and $(3, 6)$.

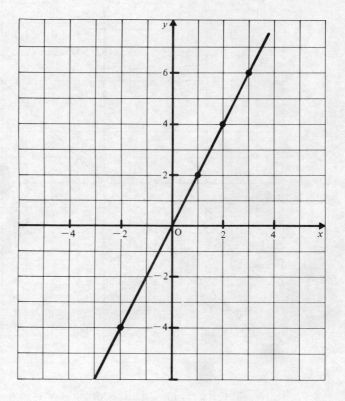

These points also lie on a straight line passing through the origin.
In each case the y-coordinate is twice the x-coordinate.
The equation of this line is therefore $y = 2x$ and we often refer, briefly, to "the line $y = 2x$".

If another point on this line has an x-coordinate of 4,
                          its y-coordinate is 2×4, i.e. 8,

and if a further point has a y-coordinate of -5,
                          its x-coordinate must be $-2\frac{1}{2}$.

EXERCISE 7a

1. Find the y-coordinates of points on the line $y = x$ which have x-coordinates of   a) 2   b) 3   c) 7   d) 12.

2. Find the y-coordinates of points on the line $y = x$ which have x-coordinates of   a) -1 b) -6 c) -8 d) -20.

3. Find the y-coordinates of points on the line $y = -x$ which have x-coordinates of   a) $3\frac{1}{2}$ b) $-4\frac{1}{2}$ c) 6.1   d) -8.3.

4. Find the x-coordinates of points on the line $y = -x$ which have y-coordinates of   a) 7   b) -2 c) $5\frac{1}{2}$ d) -4.2.

5. Find the y-coordinates of points on the line $y = 2x$ which have x-coordinates of   a) 5   b) -4 c) $3\frac{1}{2}$ d) -2.6.

6. Find the x-coordinates of points on the line $y = -3x$ which have y-coordinates of   a) 3   b) -9 c) 6   d) -4.

7. Find the x-coordinates of points on the line $y = \frac{1}{2}x$ which have y-coordinates of   a) 6   b) -12 c) $\frac{1}{2}$ d) -8.2.

8. Find the x-coordinates of points on the line $y = -4x$ which have y-coordinates of   a) 8   b) -16 c) 6   d) -3.

9. If the points $(-1, a)$ $(b, 15)$ and $(c, -20)$ lie on the straight line with equation $y = 5x$, find the values of a, b and c.

10. If the points $(3, a)$, $(-12, b)$ and $(c, -12)$ lie on the straight line with equation $y = -\frac{2}{3}x$, find the values of a, b and c.

11. Using 1 cm to 1 unit on each axis, plot the points $(-2, -6)$, $(1, 3)$, $(3, 9)$ and $(4, 12)$. What is the equation of the straight line which passes through these points?

12. Using 1 cm to 1 unit on each axis, plot the points $(-3, 6)$ $(-2, 4)$, $(1, -2)$ and $(3, -6)$. What is the equation of the straight line which passes through these points?

13. Using the same scale on each axis, plot the points $(-6, 2)$, $(0, 0)$, $(3, -1)$ and $(9, -3)$. What is the equation of the straight line which passes through these points?

14. Using the same scale on each axis, plot the points $(-6, -4)$, $(-3, -2)$, $(6, 4)$ and $(12, 8)$. What is the equation of the straight line which passes through these points?

15. Which of the points $(-2, -4)$, $(2.5, 4)$, $(6, 12)$ and $(7.5, 10)$ lie on the line $y = 2x$?

16. Which of the points $(-5, -15)$, $(-2, 6)$, $(1, -3)$ and $(8, -24)$ lie on the line $y = -3x$?

17. Which of the following points lie   a) above   b) below   the line $y = \frac{1}{2}x$: $(2, 2)$, $(-2, 1)$, $(3, 0)$, $(-4.2, -2)$, $(-6.4, -3.2)$?

PLOTTING THE GRAPH OF A GIVEN EQUATION

If we want to draw the graph of $y = 3x$ for values of x from -3 to $+3$, then we need to find the coordinates of some points on the line.

As we know that it is a straight line, two points are enough. However, it is sensible to find three points, the third point acting as a check on our working. It does not matter which three points we find, so we will choose easy values for x, one at each extreme and one near the middle.

If $x = -3$, $y = 3 \times (-3) = -9$

If $x = 0$, $y = 3 \times 0 = 0$

If $x = 3$, $y = 3 \times 3 = 9$

These look neater if we write them in table form:

x	-3	0	3
y	-9	0	9

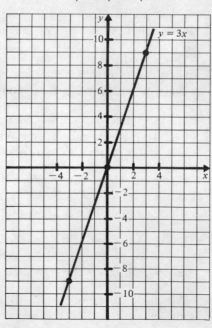

EXERCISE 7b In questions 1 to 6, draw the graphs of the given equations on the same set of axes. Use the same scale on both axes, taking values of x between -4 and 4, and values of y between -6 and 6. You should take at least three x values and record the corresponding y values in a table. Write the equation of each line somewhere on it.

1. $y = x$ **4.** $y = \frac{1}{4}x$

2. $y = 2x$ **5.** $y = \frac{1}{3}x$

3. $y = \frac{1}{2}x$ **6.** $y = \frac{3}{2}x$

In questions 7 to 12, draw the graphs of the given equations on the same set of axes.

7. $y = -x$ **10.** $y = -\frac{1}{4}x$

8. $y = -2x$ **11.** $y = -\frac{1}{3}x$

9. $y = -\frac{1}{2}x$ **12.** $y = -\frac{3}{2}x$

> We can conclude from these exercises that the graph of an equation of the form $y = mx$, is a straight line that:
> a) passes through the origin
> b) gets steeper as m increases
> c) makes an acute angle with the positive x-axis if m is positive
> d) makes an obtuse angle with the positive x-axis if m is negative.

GRADIENT OF A STRAIGHT LINE

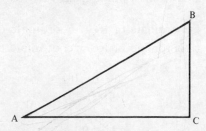

The gradient or slope of a line is defined as the amount the line rises vertically divided by the distance moved horizontally,

i.e.   gradient or slope of AB $= \dfrac{BC}{AC}$

The gradient of any line is defined in a similar way.

Considering any two points on a line, the gradient of the line is given by

$$\frac{\text{the increase in } y \text{ value}}{\text{the increase in } x \text{ value}}$$

If we plot the points O(0, 0), B(4, 4) and C(5, 5), all of which lie on the line with equation $y = x$, then:

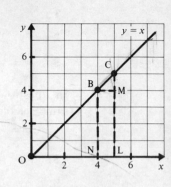

$$\text{gradient of OC} = \frac{\text{CL}}{\text{OL}} = \frac{5}{5} = 1$$

$$\text{gradient of OB} = \frac{\text{BN}}{\text{ON}} = \frac{4}{4} = 1$$

$$\text{gradient of BC} = \frac{\text{CM}}{\text{BM}} = \frac{5-4}{5-4} = \frac{1}{1} = 1$$

These show that, whichever two points are taken, the gradient of the line is 1.

Similarly, if we plot the points P(−3, 6), Q(−1, 2) and R(4, −8), all of which lie on the line with equation $y = -2x$, then:

gradient of PR

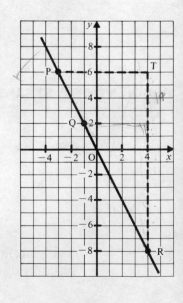

$$= \frac{\text{increase in } y \text{ value from P to R}}{\text{increase in } x \text{ value from P to R}}$$

$$= \frac{y\text{-coordinate of R} - y\text{-coordinate of P}}{x\text{-coordinate of R} - x\text{-coordinate of P}}$$

$$= \frac{(-8) - (6)}{(4) - (-3)}$$

$$= \frac{-8 - 6}{4 + 3}$$

$$= \frac{-14}{7}$$

$$= -2$$

EXERCISE 7c

Draw axes for x and y, for values between -6 and $+6$, taking 1 cm as 1 unit on each axis.

Plot the points A$(-4, 4)$ B$(2, -2)$ and C$(5, -5)$, all of which lie on the line $y = -x$. Find the gradient of

a) AB   b) BC   c) AC

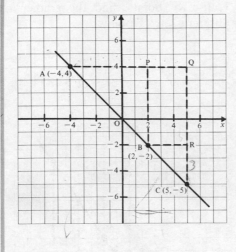

a) Gradient of AB

$$= \frac{(-2)-(4)}{(2)-(-4)} = \frac{-6}{6} = -1$$

b) Gradient of BC

$$= \frac{(-5)-(-2)}{(5)-(2)} = \frac{-3}{3} = -1$$

c) Gradient of AC

$$= \frac{(-5)-(4)}{(5)-(-4)} = \frac{-9}{9} = -1$$

1. Using 2 cm to 1 unit on each axis, draw axes which range from 0 to 6 for x and from 0 to 10 for y. Plot the points A$(2, 4)$, B$(3, 6)$ and C$(5, 10)$, all of which lie on the line $y = 2x$. Find the gradient of   a) AB   b) BC   c) AC

2. Draw the x-axis from -4 to 4 taking 2 cm as 1 unit, and the y-axis from -16 to 12 taking 0.5 cm as 1 unit. Plot the points X$(-3, 12)$, Y$(-1, 4)$ and Z$(4, -16)$, all of which lie on the line $y = -4x$. Find the gradient of a) XY   b) YZ   c) XZ

3. Choosing your own scale and range of values for both x and y, plot the points D$(-2, -6)$, E$(0, 0)$ and F$(4, 12)$, all of which lie on the line $y = 3x$. Find the gradient of   a) DE   b) EF   c) DF

4. Taking 2 cm as 1 unit for x and 1 cm as 1 unit for y, draw the x-axis from -1.5 to 2.5 and the y-axis from -10 to 6. Plot the points A$(-1.5, 6)$, B$(0.5, -2)$ and C$(2.5, -10)$, all of which lie on the line $y = -4x$. Find the gradient of   a) AB   b) BC   c) AC

Copy and complete the following table and use it to draw the graph of $y = 1.5x$.

x	-6	-4	0	2	4	6
y						

Choosing your own points, find the gradient of this line using two different sets of points.

x	-6	-4	0	2	4	6
y	-9	-6	0	3	6	9

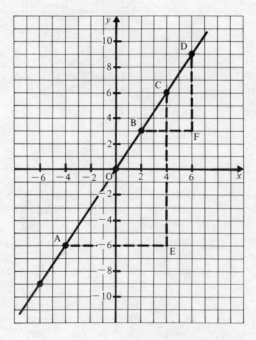

Four points, A, B, C and D, have been chosen.

$$\text{Gradient of line} = \frac{\text{CE}}{\text{AE}} = \frac{12}{8} = 1.5$$

$$\text{Gradient of line} = \frac{\text{DF}}{\text{BF}} = \frac{6}{4} = 1.5$$

(Finding the gradient using any other two points also gives a value of 1.5.)

5. Copy and complete the following table and use it to draw the graph of $y = 2.5x$.

x	-3	-1	0	2	4
y					

Choose your own pairs of points to find the gradient of this line at least twice.

6. Copy and complete the following table and use it to draw the graph of $y = -0.5x$.

x	-6	-2	3	4
y				

Choose your own pairs of points to find the gradient of this line at least twice.

7. Determine whether the straight lines with the following equations have positive or negative gradients:

a) $y = 5x$ d) $y = -\frac{1}{4}x$

b) $y = -7x$ e) $3y = -x$

c) $y = 12x$ f) $5y = 12x$

These exercises, together with the worked examples, confirm our conclusions on p. 99, namely that
a) the larger the value of m the steeper is the slope
b) lines with positive values for m make an acute angle with the positive x-axis
c) lines with negative values for m make an obtuse angle with the positive x-axis.

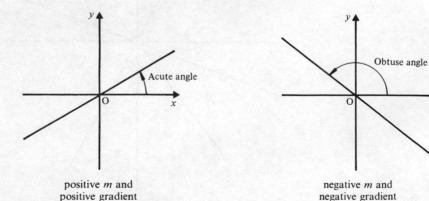

positive m and   negative m and
positive gradient   negative gradient

EXERCISE 7d   For each of the following pairs of lines, state which line is the steeper. Show both lines on the same sketch.

1. $y = 5x, \quad y = \frac{1}{5}x$ **5.** $y = 10x, \quad y = 7x$

2. $y = 2x, \quad y = 5x$ **6.** $y = -\frac{1}{2}x, \quad y = -\frac{1}{4}x$

3. $y = \frac{1}{2}x, \quad y = \frac{1}{3}x$ **7.** $y = -6x, \quad y = -3x$

4. $y = -2x, \quad y = -3x$ **8.** $y = 0.5x, \quad y = 0.75x$

Determine whether each of the following straight lines makes an acute angle or an obtuse angle with the positive x-axis.

9. $y = 4x$ **15.** $y = 10x$

10.. $y = -3x$ **16.** $y = 0.5x$

11. $y = -\frac{1}{2}x$ **17.** $y = -6x$

12. $y = 3.6x$ **18.** $y = -\frac{2}{3}x$

13. $y = \frac{1}{3}x$ **19.** $y = -\frac{3}{4}x$

14. $y = 0.7x$ **20.** $y = -0.4x$

21.   Estimate the gradient of each of the lines shown in the sketch.

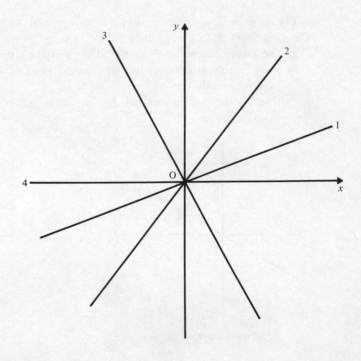

LINES THAT DO NOT PASS THROUGH THE ORIGIN

If we plot the points $(-3, -1)$, $(1, 3)$, $(3, 5)$, $(4, 6)$ and $(6, 8)$, and draw the straight line that passes through these points, we can use it to find

a) the equation of the line
b) its gradient
c) the distance from the origin to the point where the line crosses the y-axis.

a) In each case, the y-coordinate is 2 more than the x-coordinate, i.e. all the points lie on the line with equation $y = x + 2$

b) Using the points A and B, the gradient of the line is given by $\dfrac{BC}{AC}$, i.e. $\dfrac{3}{3} = 1$.

c) The line crosses the y-axis at the point $(0, 2)$ which is 2 units above the origin. This quantity is called the *y intercept*.

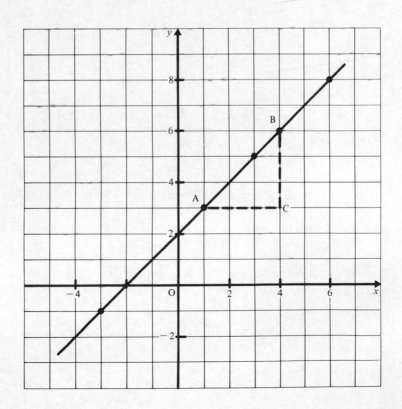

EXERCISE 7e

Draw the graph of $y = -4x + 3$ for values of x between -4 and $+4$. Hence find a) the gradient of the line b) its y intercept.

x	-3	0	3
y	15	3	-9

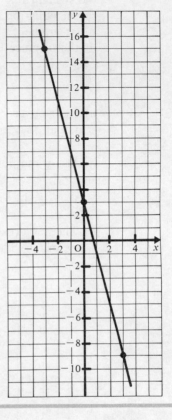

a) Gradient of line $= \dfrac{-4}{1}$

$\qquad\qquad\qquad = -4$

b) The y intercept is 3

In the following questions, draw the graph of the given equation using the given x values. Hence find the gradient of the line and its intercept on the y-axis. Use 1 cm as 1 unit on each axis with x values ranging from -8 to $+8$ and y values ranging from -10 to $+10$.

Compare the values you get for the gradient and the y intercept with the numbers in the right-hand side of each equation.

1. $y = 3x + 1$; x values $\quad -3, 1, 3$
Use your graph to find the value of y when x is a) -2 b) 2

2. $y = -3x + 4$; x values $\quad -2, 2, 4$
Use your graph to find the value of y when x is a) -1 b) 3

3. $y = \frac{1}{2}x + 4$; x values $-8, 0, 6$
Use your graph to find a) the value of y when x is -2
b) the value of x when y is 6

4. $y = x - 3$; x values $-4, 2, 8$
Use your graph to find the value of x when y is a) 4 b) -5

5. $y = \frac{3}{4}x + 3$; x values $-4, 0, 8$
Use your graph to find the value of x when y is a) 6 b) 4.5

Draw the graph of $y = -2x + 3$ for values of x between
-4 and $+4$. Hence find
a) the gradient of the line   b) its y intercept.

Compare the values for the gradient and the y intercept
with the number of xs and the number term on the
right-hand side of the equation.

x	-4	0	4
y	11	3	-5

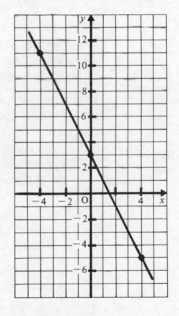

a)   The gradient $= -\dfrac{8}{4}$

$= -2$

b)   The y intercept is 3

The number of xs on the right-hand side is -2, which is
the same as the gradient of the line.

The number term on the right-hand side is 3, which is the
same as the y intercept.

In the following questions, draw a graph for each of the given equations. In each case find the gradient and the y intercept for the resulting straight line. Take 1 cm as 1 unit on each axis, together with suitable values of x within the range -4 to $+4$. Choose your own range for y when you have completed the table.

Compare the values you get for the gradient and the y intercept with
a) the number of xs
b) the number term on the right-hand side of the equation.

6. $y = 2x-2$ **11.** $y = 2x+5$

7. $y = -2x+4$ **12.** $y = -2x-7$

8. $y = 3x-4$ **13.** $y = -3x+2$

9. $y = \frac{1}{2}x+3$ **14.** $y = \frac{1}{3}x-6$

10. $y = -\frac{3}{2}x+3$ **15.** $y = \frac{2}{5}x-5$

THE EQUATION $y = mx+c$

The results of Exercise 7e show that we can "read" the gradient and the y intercept of a straight line from its equation.
For example, the line with equation $y = 3x-4$ has a gradient of 3 and its y intercept is -4.

In general we can conclude that the equation $y = mx+c$ gives a straight line where m is the gradient of the line and c is the y intercept.

EXERCISE 7f

> Write down the gradient, m, and the y intercept, c, for the straight line with equation $y = 5x-2$
>
> For the line $y = 5x-2$
>
> $$m = 5 \quad \text{and} \quad c = -2$$

Write down the gradient, m, and y intercept, c, for the straight line with the given equation.

1. $y = 4x+7$ **6.** $y = \frac{2}{5}x-3$

2. $y = \frac{1}{2}x-4$ **7.** $y = \frac{3}{4}x+7$

3. $y = 3x-2$ **8.** $y = 4-3x$

4. $y = -4x+5$ **9.** $y = 6-\frac{1}{2}x$

5. $y = 7x+6$ **10.** $y = -3-7x$

Sketch the straight line with equation $y = 5x - 7$

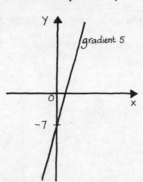

Sketch the straight lines with the given equations.

11. $y = 2x + 5$

12. $y = 7x - 2$

13. $y = \frac{1}{2}x + 6$

14. $y = -2x - 3$

15. $y = -\frac{2}{3}x + 8$

16. $y = 4x + 2$

17. $y = -5x - 3$

18. $y = 3x + 7$

19. $y = \frac{3}{4}x - 2$

20. $y = \frac{1}{3}x - 5$

Sketch the straight line with equation $y = 2 - 3x$

First rearrange the equation in the form $y = mx + c$

i.e. $y = -3x + 2$

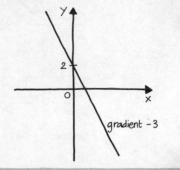

21. $y = 4 - x$

22. $y = 3 - 2x$

23. $y = 8 - 4x$

24. $y = -3 - x$

25. $y = 2(x + 1)$

26. $y = 3(x - 2)$

27. $y = -5(x - 1)$

28. $y = 3(4 - x)$

29. $y = -2(2x + 3)$

30. $y = -3(x - 4)$

PARALLEL LINES

Lines with the same gradient are said to be parallel.

The diagram shows the lines $y = x+2$ and $y = x-3$.

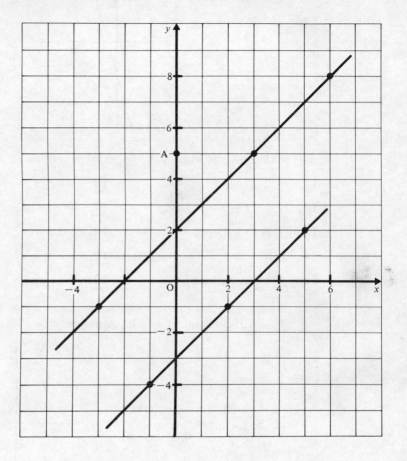

These lines have the same gradient, i.e. they are parallel.

Now consider a third line, parallel to the first two lines and passing through the point A(0,5).

Its gradient is the same as that of the first lines, i.e. $m = 1$.

It crosses the y-axis at (0,5) so its y intercept is 5, i.e. $c = 5$.

Therefore the equation of the third line is $y = x+5$.

Similarly the equation of another parallel line passing through the point $(0, -5)$ is $y = x-5$.

EXERCISE 7g **1.** Draw the graphs of $y = 3x+1$ and $y = 3x-4$ taking x values of -2, 2 and 3.
(Let x range from -5 to $+5$ and y from -10 to $+10$.
Take 1 cm to represent 1 unit on each axis.)

What do you notice about these lines?
What do you notice about their m values?

2. Draw the graphs of $y = -2x+3$ and $y = -2x-3$ taking x values of -3, 0 and 3.
(Take 1 cm to represent 1 unit on each axis.
Let x range from -6 to $+6$ and y from -10 to $+10$.)

What do you notice about these lines?
What do you notice about their m values?

By finding the gradient of each line, determine whether or not the given pairs of equations represent parallel lines.

3. $y = 4x+2$, $y = 4x-7$ **7.** $y = -x+4$, $y = -x-3$

4. $y = \frac{1}{2}x+6$, $y = \frac{1}{2}x+10$ **8.** $y = -5x+2$, $y = -5x-13$

5. $y = x+4$, $y = 2x+4$ **9.** $y = \frac{2}{3}x+3$, $y = \frac{1}{3}x-4$

6. $y = 3x+5$, $y = x+7$ **10.** $y = \frac{1}{2}x-4$, $y = 0.5x+2$

Find the gradient of each of the lines $x+y = 4$ and $y = -x+2$. Hence determine whether or not the two lines are parallel.

$x+y = 4$ (1)
$y = -x+2$ (2)

Equation (1) gives $y = -x+4$
the gradient of this line is -1

The gradient of the line $y = -x+2$ is -1
i.e. the lines have the same gradient and are therefore parallel.

Find the gradient of each of the lines in each question. Hence determine whether or not the two lines are parallel.

11. $y = 2x+3$, $2y = 4x-7$ **14.** $3y = 5x+7$, $6y = 10x-3$

12. $3y = 9x-2$, $y = 3x+13$ **15.** $5y = x+2$, $3y = x+2$

13. $x+y = 5$, $y = -2x+3$ **16.** $x+y = 4$, $y = -x+6$

LINES PARALLEL TO THE AXES

We began by considering the equation $y = mx$,
i.e. the equation $y = mx + c$ when $c = 0$.
This equation gave a straight line passing through the origin.
Now we will see what happens when $m = 0$.
Think, for example, of the equation $y = 3$.
For every value of x the y-coordinate is 3. This means that the graph of $y = 3$ is a straight line parallel to the x-axis at a distance 3 units above it.

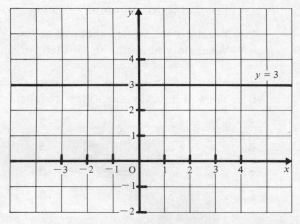

$y = c$ is therefore the equation of a straight line parallel to the x-axis at a distance c away from it. If c is positive, the line is above the x-axis, and if c is negative, the line is below the x-axis.

Similarly $x = b$ is the equation of a straight line parallel to the y-axis at a distance b units from it.

EXERCISE 7h

Draw, on the same diagram, the straight line graphs of $x = -3$, $x = 5$, $y = -2$ and $y = 4$.

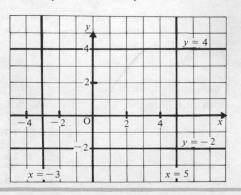

In the following questions, take both x and y in the range -8 to $+10$. Let 1 cm be 1 unit on each axis.

1. Draw the straight line graphs of the following equations in a single diagram: $x = 2$, $x = -5$, $y = \frac{1}{2}$, $y = -3\frac{1}{2}$

2. Draw the straight line graphs of the following equations in a single diagram: $y = -5$, $x = -3$, $x = 6$, $y = 5.5$

3. On one diagram, draw graphs to show the following equations:
   $x = 5$, $y = -5$, $y = 2x$
   Write down the coordinates of the three points where these lines intersect. What kind of triangle do they form?

4. On one diagram, draw the graphs of the straight lines with equations $x = 4$, $y = -\frac{1}{2}x$, $y = 3$
   Write down the coordinates of the three points where these lines intersect. What kind of triangle is it?

5. On one diagram, draw the graphs of the straight lines with equations $y = 2x+4$, $y = -5$, $y = 4-2x$
   Write down the coordinates of the three points where these lines intersect. What kind of triangle is it?

MIXED EXERCISES

EXERCISE 7i

1. Find the x-coordinates of the points on the line $y = 3x$ that have y-coordinates of   a) 6   b) -12 c) 2.

2. If the points $(6, a)$, $(-\frac{1}{2}, b)$ and $(c, 1)$ lie on the straight line with equation $3y = -2x$, find the values of a, b and c.

3. Determine whether the straight lines with the given equations have positive or negative gradients:
   a) $y = 4x$
   b) $y = -2x+2$
   c) $y = \frac{2}{3}x - 7$

4. Copy and complete the following table and use it to draw the graph of $y = 2x-3$:

x	-3	0	4
y			

   Choose your own points to find the gradient of this line.

5. Determine in each case whether the straight line with the given equation makes an acute angle or an obtuse angle with the positive x-axis.

a) $y = -\frac{2}{3}x$

b) $y = 5x + 2$

c) $2x + y = 3$

d) $3y = -4x + 7$

6. Draw on the same axes, using 1 cm as 1 unit in each case, the graphs of $y = 2x - 4$ and $2x + y + 8 = 0$. Write down the coordinates of the point where these lines intersect.

EXERCISE 7j

1. Find the y-coordinates of the points on the line $y = 5x$ that have x-coordinates of   a) 2   b) 3   c) $\frac{1}{2}$.

2. If the points $(-1, a)$, $(b, 15)$ and $(c, -20)$ lie on the straight line with equation $y = 5x$, find the values of a, b and c.

3. Determine whether the straight lines with the given equations have positive or negative gradients:

a) $y = 6x$

b) $y = -3x + 2$

c) $x + y = 4$

4. Write down the gradients and y intercepts for the straight lines with the given equations:

a) $y = 4x - 7$

b) $2y = 5x + 2$

c) $y - 3x = 2$

d) $3y = -x - 12$

5. Determine whether or not the given pairs of equations represent parallel lines:

a) $y = -x + 2$, $x + y = 3$

b) $2y = 4x + 3$, $y + 2x = 5$

6. Draw, on the same axes, the graphs of $x = -3$, $y = \frac{1}{2}x$ and $y = 4$, for values of x between -4 and $+8$. Write down the coordinates of the three points where these lines intersect.

EXERCISE 7k

1. Find the y-coordinates of the points on the line $y = 7x + 4$ that have x-coordinates of   a) 1   b) -2 c) -5.

2. If the points $(3, a)$, $(-2, b)$ and $(c, -10)$ lie on the straight line with equation $y = 5 - 3x$, find the values of a, b and c.

3. Sketch on the same axes the graphs of the straight lines with equations   a) $y = -3x$ b) $y = 2x + 4$.

4. Draw the graph of $y = 5x - 2$ for values of x between -4 and 4. Use 2 cm as 1 unit on the x-axis and 1 cm as 1 unit on the y-axis. From your graph, or otherwise, find

   a) the gradient of the line   b) its y-intercept.

5. Write down the equations of the straight lines that have the given gradients and y intercepts:

   a) gradient 2, y intercept -4

   b) gradient $\frac{1}{2}$, y intercept 5

   c) gradient -4, y intercept -3

6. Draw, on the same axes, the graphs of $x = 1$, $y = -2x - 2$, $y = 4$ for values of x between -4 and $+4$. Write down the coordinates of the three points where these lines intersect.

8 REFLECTIONS AND TRANSLATIONS

LINE SYMMETRY

 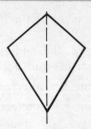 

As we saw in Book 1, shapes like these are *symmetrical*. They have line symmetry (or bilateral symmetry); the dotted line is the *axis of symmetry* because if the shape were folded along the dotted line, one half of the drawing would fit exactly over the other half.

EXERCISE 8a   **1.**   Which of the following shapes have an axis of symmetry?

a)    b)   c)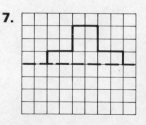

Copy the following drawings on squared paper and complete them so that the dotted line is the axis of symmetry.

2.    **4.** 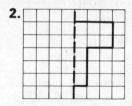   **6.**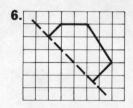

3.    **5.** 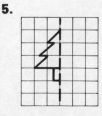   **7.**

TWO OR MORE AXES OF SYMMETRY

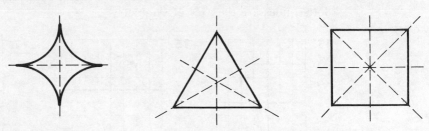

Shapes can have more than one axis of symmetry. In the drawings above, the axes are shown by dotted lines and it is clear that the first shape has two axes of symmetry, the second has three and the third has four.

EXERCISE 8b   Sketch or trace the shapes in questions 1 to 12 and mark in the axes of symmetry. (Some shapes may have no axis of symmetry.)

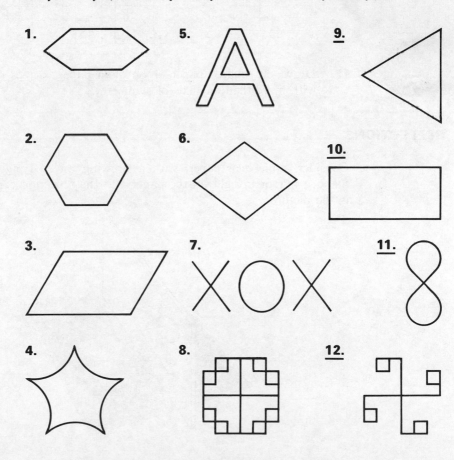

Copy and complete the following drawings on squared paper. The dotted lines are the axes of symmetry.

13. 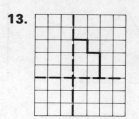   **15.** 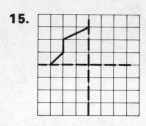   **17.**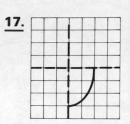

14. 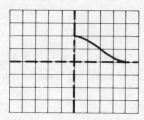   **16.** 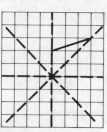   **18.**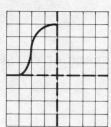

19. Draw, on squared paper or on plain paper, shapes of your own with more than one axis of symmetry.

REFLECTIONS

Consider a piece of paper, with a drawing on it, lying on a table. Stand a mirror upright on the paper and the reflection can be seen as in the picture.

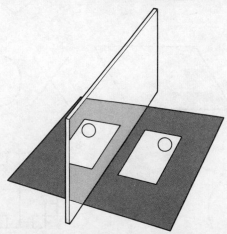

If we did not know about such things as mirrors, we might imagine that there were two pieces of paper lying on the table like this:

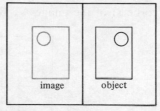

image   object

The *object* and the *image* together form a symmetrical shape and the *mirror line* is the axis of symmetry.

EXERCISE 8c   In this exercise it may be helpful to use a small rectangular mirror, or you can use tracing paper to trace the object and turn the tracing paper over, to find the shape of the image.

Copy the objects and mirror lines (indicated by dotted lines) on to squared paper and draw the image of each object.

1.

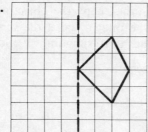

4.

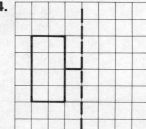

2.

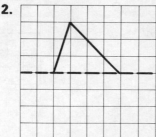

5.

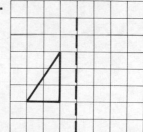

3.

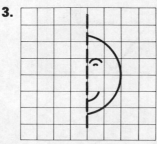

6.

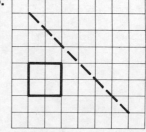

Copy triangle ABC and the mirror line on to squared paper. Draw the image. Label the corresponding vertices (corners) of the image A′, B′, C′.

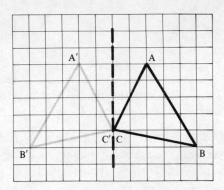

(In this case C and C′ are the same point.)

In each of the following questions, copy the object and the mirror line on to squared paper. Draw the image. Label the vertices of the object A, B, C, etc. and label the corresponding vertices of the image A′, B′, C′, etc.

7.

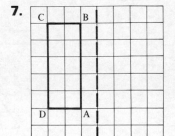

8.

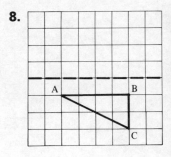

9.

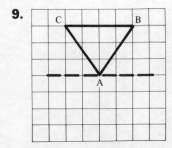

In mathematical reflection, though not in real life, the object can cross the mirror line.

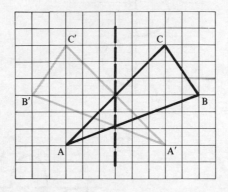

10.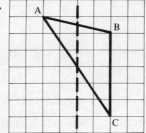

11.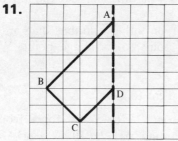

12.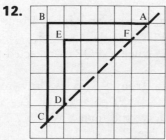

13.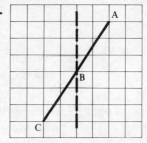

14.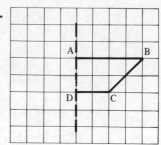

15.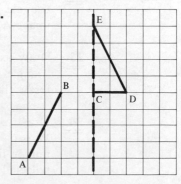

16.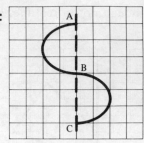

17.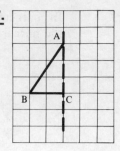

18.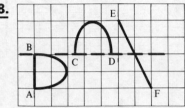

19. Which points in questions 7 to 18 are labelled twice? What is special about their positions?

20. In the diagram for question 10, join A and A′.

a) Measure the distances of A and A′ from the mirror line. What do you notice?

b) At what angle does the line AA′ cut the mirror line?

21. Repeat question 20 on other suitable diagrams, in each case joining each object point to its image point. What conclusions do you draw?

In questions 22 to 25 use 1 cm to 1 unit.

22. Draw axes, for x from -5 to 5 and for y from 0 to 5. Draw triangle ABC by plotting A(1, 2), B(3, 2) and C(3, 5). Draw the image A′B′C′ when ABC is reflected in the y-axis.

23. Draw axes, for x from 0 to 5 and for y from -2 to 2. Draw triangle PQR where P is $(1, -1)$, Q is $(5, -1)$ and R is $(4, 0)$. Draw the image P′Q′R′ when $\triangle$PQR is reflected in the x-axis.

24. Draw axes for x and y from -5 to 1. Draw rectangle WXYZ: W is $(-3, -1)$, X$(-3, -2)$, Y$(-5, -2)$ and Z$(-5, -1)$. Draw the mirror line $y = x$. Draw the image W′X′Y′Z′ when WXYZ is reflected in the mirror line.

25. Draw axes for x and y from -1 to 9. Plot the points A(2, 1), B(5, 1), C(7, 3) and D(4, 3). Draw the parallelogram ABCD and its image by reflection in the line $y = x$.

26. Draw axes for x and y from -6 to 8. Draw triangle ABC where A is $(-6, -2)$, B is $(-3, -4)$ and C is $(-2, -1)$. Draw the following images of triangle ABC:

a) triangle $A_1B_1C_1$ by reflection in the y-axis

b) triangle $A_2B_2C_2$ by reflection in the line $y = -x$ (this is the straight line through the points $(2, -2)$, $(-4, 4)$)

c) triangle $A_3B_3C_3$ by reflection in the x-axis

d) triangle $A_4B_4C_4$ by reflection in the line $x = -1$

INVARIANT POINTS

A point which is its own image, i.e. such that the object point and its image are in the same place, is called an *invariant point*. The previous examples showed that, with reflection, the invariant points lie on the mirror line. The mirror line is an *invariant line*.

FINDING THE MIRROR LINE

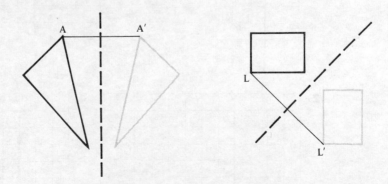

We can see from these diagrams, and from the work in the previous exercise, that the object and the image points are at equal distances from the mirror line, and the lines joining them (e.g. AA′ and LL′) are perpendicular (at right angles) to the mirror line.

EXERCISE 8d

Find the mirror line if △A′B′C′ is the image of △ABC.

(The mirror line is halfway between A and A′ and perpendicular to AA′.)

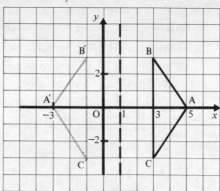

The mirror line is the line $x = 1$

Copy the diagrams in questions 1 to 4 and draw in the mirror lines.

1.

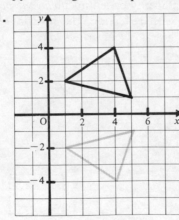

3.

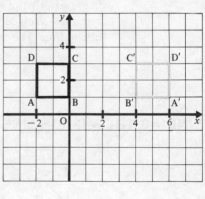

2.

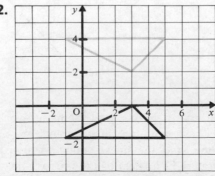

4.

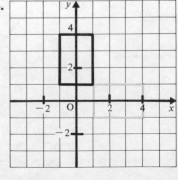

Draw axes for x and y from -5 to 5 for each of questions 5 to 8.

5. Draw square PQRS: P(1, 1), Q(4, 1), R(4, 4), S(1, 4). Draw square P'Q'R'S': P'(−2, 1), Q'(−5, 1), R'(−5, 4), S'(−2, 4). Draw the mirror line so that P'Q'R'S' is the reflection of PQRS and write down its equation.

6. Draw △XYZ: X(2, 1), Y(4, 4), Z(−2, 4), and △X'Y'Z': X'(2, 1), Y'(4, −2), Z'(−2, −2). Draw the mirror line so that △X'Y'Z' is the reflection of △XYZ and write down its equation. Are there any invariant points? If there are, name them.

7. Draw △ABC: A(−2, 0), B(0, 2), C(−3, 3), and △PQR: P(3, −1), Q(4, −4), R(1, −3). Draw the mirror line so that △PQR is the reflection of △ABC. Which point is the image of A? Are there any invariant points? If there are, name them.

8. Draw lines AB and PQ: A(2, −1), B(4, 4), P(−2, −1), Q(−5, 4). Is PQ a reflection of AB? If it is, draw the mirror line. If not, give a reason.

If A'B'C' is the reflection of ABC, draw the mirror line.

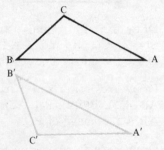

(Join AA' and BB' and find the midpoints P and Q. Then PQ is the mirror line.)

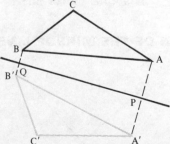

Whenever you attempt to draw a mirror line in this way, always check that the mirror line is at right angles to AA' and BB'. If it is not, then A'B'C' cannot be a reflection of ABC.

9. Trace the diagrams and draw the mirror lines.

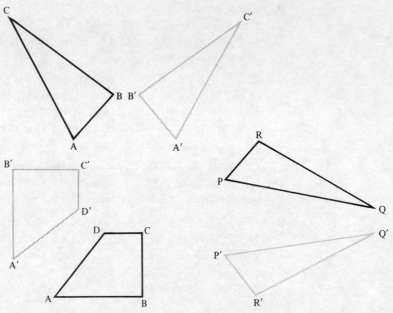

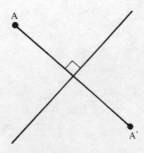

10. Draw axes for x and y from -4 to 5. Draw $\triangle ABC$: A(3, 1), B(4, 5), C(1, 4), and $\triangle A'B'C'$: A'(0, -2), B'(-4, -3), C'(-3, 0). Draw the mirror line so that $\triangle A'B'C'$ is the image of $\triangle ABC$.

11. Draw axes for x and y from -4 to 4. Draw lines AB and PQ: A(-4, 3), B(0, 4), P(1, -2), Q(2, 2). Draw the mirror line so that AB is the image of PQ.

12. Draw axes for x and y from -3 to 5. Draw $\triangle XYZ$: X(3, 2), Y(5, 2), Z(3, 5), and $\triangle LMN$: L(0, -3), M(0, -1), N(-3, -1). Draw the mirror line.

CONSTRUCTION OF THE MIRROR LINE

If we have only one point and its image, and we cannot use squares to guide us, we can use the fact that the mirror line goes through the midpoint of AA' and is perpendicular to AA'. The mirror line is therefore the perpendicular bisector of AA' and can be constructed with compasses.

EXERCISE 8e **1.** On plain paper mark two points P and P′ about 10 cm apart in the middle of the page and construct the perpendicular bisector of PP′. Join PP′ and check that it is cut in half by the line you have constructed and that the two lines cut at right angles. Are we correct in saying that P′ is the reflection of P in the constructed line?

2. On squared paper draw axes for x and y from -5 to 5, using 1 cm to 1 unit. A is the point $(5, 2)$ and A′ is the point $(-3, -3)$. Construct the mirror line so that A′ is the reflection of A.

3. Draw axes for x and y from -1 to 8, using 1 cm to 1 unit. B is the point $(-1, 0)$ and B′ is the point $(6, 3)$. Construct the mirror line so that B′ is the reflection of B.

4. Find the gradient and the y intercept of the mirror line in question 3. Hence find the equation of the mirror line.

OTHER TRANSFORMATIONS

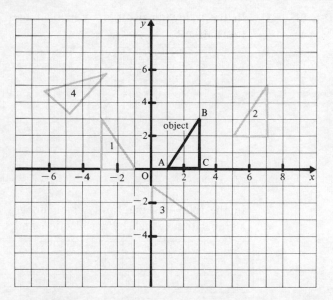

Imagine a triangle ABC cut out of card and lying in the position shown. We can reflect △ABC in the y-axis by picking up the card, turning it over and putting it down again in position 1.

Starting again from its original position, we can change its position by sliding the card over the surface of the paper to position 2, 3 or 4. Some of these movements can be described in a simple way, some are more complicated.

TRANSLATIONS

Consider the movements in the diagram:

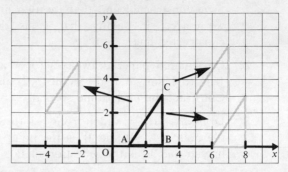

All these movements are of the same type. The side AB remains parallel to the *x*-axis in each case and the triangle continues to face in the same direction. This type of movement is called a *translation*.

Although not a reflection we still use the words *object* and *image*.

EXERCISE 8f   **1.** In the following diagram, which images of △ABC are given by translations?

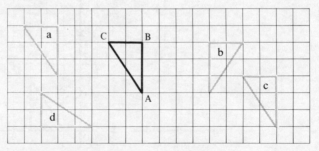

2. In the following diagram, which images of △ABC are given by a translation, which by a reflection and which by neither?

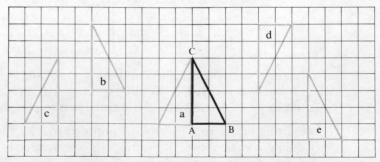

3. Repeat question 2 with the diagram on page 127.

DESCRIPTIONS OF TRANSLATIONS

EXERCISE 8g Draw sketches to illustrate the following translations:

1. An object is translated 6 cm to the left.

2. An object is translated 4 units parallel to the *x*-axis to the right.

3. An object is translated 3 m due north.

4. An object is translated 5 km south-east.

5. An object is translated 3 units parallel to the *x*-axis to the right and then 4 units parallel to the *y*-axis upwards.

USING VECTORS TO DESCRIBE TRANSLATIONS

The translation in question 5, Exercise 8g, can be given more briefly in vector form as $\begin{pmatrix} 3 \\ 4 \end{pmatrix}$.

In Book 1 we saw that the top number gives the displacement parallel to the *x*-axis and the lower number gives the displacement parallel to the *y*-axis.

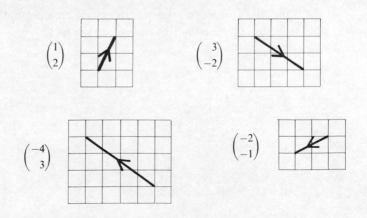

If the top number is negative, the displacement is to the left and if the lower number is negative, the displacement is downwards.

Consider the diagram:

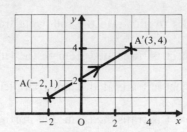

$$\overrightarrow{AA'} = \begin{pmatrix} 5 \\ 3 \end{pmatrix}$$

A' is the image of A under the translation described by the vector $\begin{pmatrix} 5 \\ 3 \end{pmatrix}$.

A is *mapped* to A' by the translation described by the vector $\begin{pmatrix} 5 \\ 3 \end{pmatrix}$.

EXERCISE 8h

A is the point $(1, 2)$. Find the image of A under the translation described by the vector $\begin{pmatrix} -4 \\ 2 \end{pmatrix}$.

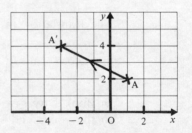

The image of A is $(-3, 4)$.

Find the images of the points given in questions 1 to 10 under the translations described by the given vectors.

1. $(3, 1)$, $\begin{pmatrix} 4 \\ 2 \end{pmatrix}$

6. $(4, -4)$, $\begin{pmatrix} 2 \\ -3 \end{pmatrix}$

2. $(4, 5)$, $\begin{pmatrix} 2 \\ 4 \end{pmatrix}$

7. $(-6, -3)$, $\begin{pmatrix} 4 \\ 1 \end{pmatrix}$

3. $(-2, 4)$, $\begin{pmatrix} 4 \\ 3 \end{pmatrix}$

8. $(1, 1)$, $\begin{pmatrix} -5 \\ -3 \end{pmatrix}$

4. $(3, 2)$, $\begin{pmatrix} -2 \\ 3 \end{pmatrix}$

9. $(3, -2)$, $\begin{pmatrix} 6 \\ -4 \end{pmatrix}$

5. $(4, 5)$, $\begin{pmatrix} -3 \\ -2 \end{pmatrix}$

10. $(7, 4)$, $\begin{pmatrix} -5 \\ -4 \end{pmatrix}$

In questions 11 to 20, find the vectors describing the translations that map A to A′.

11. A(1, 2), A′(5, 3)

12. A(3, 8), A′(2, 9)

13. A(1, 2), A′(5, 4)

14. A(−3, 0), A′(4, 6)

15. A(−4, −3), A′(0, 0)

16. A(−2, 6), A′(2, 6)

17. A(6, 9), A′(2, 3)

18. A(4, 8), A′(1, 9)

19. A(−3, −4), A′(−5, −6)

20. A(4, −2), A′(5, −1)

In questions 21 to 26, the given point A′ is the image of an object point A under the translation described by the given vector. Find A.

21. A′(7, 9), $\begin{pmatrix} 2 \\ 3 \end{pmatrix}$

22. A′(3, 6), $\begin{pmatrix} 1 \\ 4 \end{pmatrix}$

23. A′(0, 6), $\begin{pmatrix} 2 \\ 3 \end{pmatrix}$

24. A′(1, 2), $\begin{pmatrix} -2 \\ -3 \end{pmatrix}$

25. A′(6, 3), $\begin{pmatrix} -3 \\ 2 \end{pmatrix}$

26. A′(−3, −2), $\begin{pmatrix} 1 \\ 3 \end{pmatrix}$

A translation moves each point of an object the same distance in the same direction.

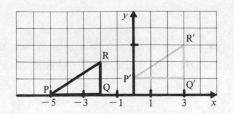

$$\overrightarrow{PP'} = \begin{pmatrix} 5 \\ 1 \end{pmatrix} \qquad \overrightarrow{RR'} = \begin{pmatrix} 5 \\ 1 \end{pmatrix}$$

$$\overrightarrow{QQ'} = \begin{pmatrix} 5 \\ 1 \end{pmatrix}$$

i.e. $\overrightarrow{PP'} = \overrightarrow{QQ'} = \overrightarrow{RR'}$

EXERCISE 8i   **1.** Given the following diagrams, find the vectors $\overrightarrow{AA'}$, $\overrightarrow{BB'}$ and $\overrightarrow{CC'}$. Are they all equal? Is the transformation a translation?

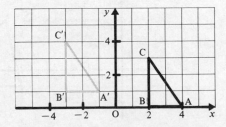

2. Given the following diagrams, find the vectors $\overrightarrow{LL'}$, $\overrightarrow{MM'}$ and $\overrightarrow{NN'}$. Are they all equal? Is the transformation a translation?

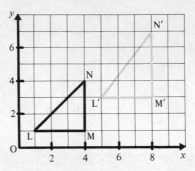

3. Find the vector which describes the translation mapping A to A', B to B' and C to C'.

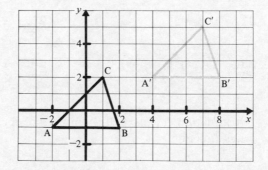

4. Give the vectors describing the translations which map
a) △ABC to △PQR   b) △PQR to △ABC.

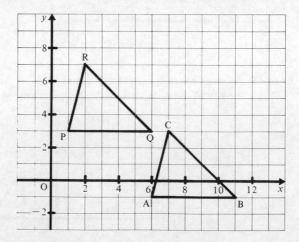

5. Give the vectors describing the translations which map
   a) △ABC to △PQR   c) △XYZ to △ABC
   b) △ABC to △LMN   d) △ABC to △ABC

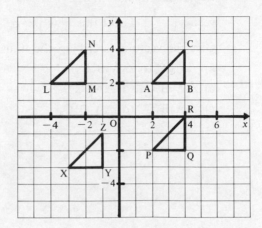

6. Draw axes for x and y from -4 to 5. Draw the following triangles: △ABC with A(2, 2), B(4, 2), C(2, 5);
   △PQR with P(1, -2), Q(3, -2), R(1, 1);
   △XYZ with X(-3, 1), Y(-1, 1), Z(-3, 4).
   Give the vectors describing the translations which map
   a) △ABC to △PQR   c) △PQR to △XYZ
   b) △PQR to △ABC   d) △ABC to △ABC

7. Draw axes for x and y from 0 to 9. Draw △ABC with A(3, 0), B(3, 3), C(0, 3) and △A′B′C′ with A′(8, 2), B′(8, 5), C′(5, 5).
   Is △A′B′C′ the image of △ABC under a translation? If so, what is the vector describing the translation?
   Join AA′, BB′ and CC′. What type of quadrilateral is AA′B′B? Give reasons. Name other quadrilaterals of the same type in the figure.

8.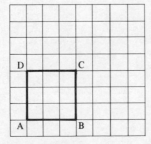

   a) Square ABCD is translated parallel to AB a distance equal to AB. Sketch the diagram and draw the image of ABCD.
   b) Square ABCD is translated parallel to AC a distance equal to AC. Sketch the diagram and draw the image of ABCD.

9. Draw axes for x and y from -2 to 7. Draw $\triangle ABC$ with $A(-2, 5)$, $B(1, 3)$, $C(1, 5)$. Translate $\triangle ABC$ using the vector $\begin{pmatrix} 5 \\ 1 \end{pmatrix}$. Label this image $A_1 B_1 C_1$. Then translate $\triangle A_1 B_1 C_1$ using the vector $\begin{pmatrix} -1 \\ -3 \end{pmatrix}$. Label this new image $A_2 B_2 C_2$.

Give the vectors describing the translations which map

a) $\triangle ABC$ to $\triangle A_2 B_2 C_2$

b) $\triangle A_2 B_2 C_2$ to $\triangle ABC$

c) $\triangle A_2 B_2 C_2$ to $\triangle A_1 B_1 C_1$

9 ROTATIONS

ROTATIONAL SYMMETRY

Some shapes have a type of symmetry different from line symmetry.

a) 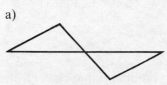 b) 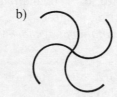 c)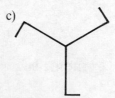

These shapes do not have an axis of symmetry but can be turned or rotated about a centre point and still look the same.

EXERCISE 9a **1.** a) 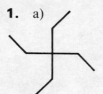 b) 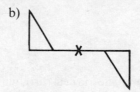 c)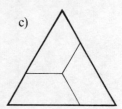

Trace each of the shapes above, then turn the tracing paper about the centre of rotation (put a compass point or a pencil point in the centre). Turn until the traced shape fits over the original shape again. In each case state through what fraction of a complete turn the shape has been rotated.

2. Which of the following shapes have rotational symmetry?

a) 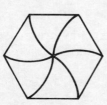 c)

b)  d)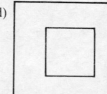

ORDER OF ROTATIONAL SYMMETRY

If a shape needs to be turned through a third of a complete turn to fit, then it will need two more such turns to return it to its original position. So, starting from its original position, it takes three turns, each one-third of a revolution, to return it to its starting position.

It has *rotational symmetry of order 3*.

EXERCISE 9b

Give the order of rotational symmetry of the following shape.

The angle turned through is a right angle or one-quarter of a complete turn.

The shape has rotational symmetry of order 4.

1. Give the orders of rotational symmetry of the shapes in Exercise 9a, question 1.

2. Give the orders of rotational symmetry, if any, of the shapes in Exercise 9a, question 2.

Copy and complete the diagram given that there is rotational symmetry of order 4

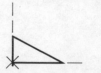

Each of the diagrams in questions 3 to 8 has rotational symmetry of
the order given and × marks the centre of rotation.
Copy and complete the diagrams. (Tracing paper may be helpful.)

3. 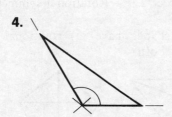   Rotational symmetry of order 4

4.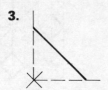

Rotational symmetry of order 3

5. 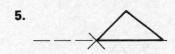   Rotational symmetry of order 2

6.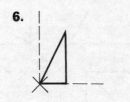

Rotational symmetry of order 4

7.

Rotational symmetry of order 3

8.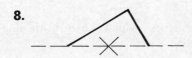

Rotational symmetry of order 2

9. In questions 3 to 8, give the size of the angle, in degrees,
through which each shape is turned.

EXERCISE 9c   Some shapes have both line symmetry and rotational symmetry:

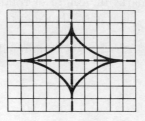 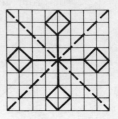

Two axes of symmetry   Four axes of symmetry
Rotational symmetry order 2   Rotational symmetry of order 4

Which of the following shapes have   a) rotational symmetry only
b) line symmetry only   c) both?

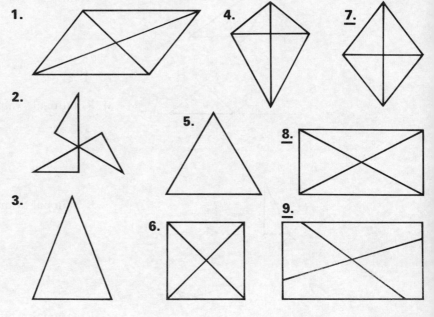

10. Make up three shapes which have rotational symmetry only. Give the order of symmetry and the angle of turn, in degrees.

11. Make up three shapes with line symmetry only. Give the number of axes of symmetry.

12. Make up three shapes which have both line symmetry and rotational symmetry.

13. The capital letter **X** has line symmetry (two axes) and rotational symmetry (of order 2). Investigate the other letters of the alphabet.

TRANSFORMATIONS: ROTATIONS

a)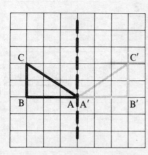

b)

c)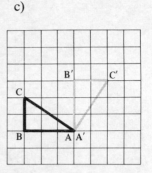

So far, in transforming an object we have used reflections, as in (a), and translations, as in (b), but for (c) we need a rotation.

In this case we are rotating △ABC about A through 90° clockwise (↷). We could also say △ABC was rotated through 270° anticlockwise (↶).

For a rotation of 180° we do not need to say whether it is clockwise or anticlockwise.

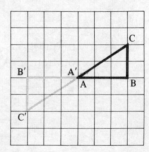

EXERCISE 9d

Give the angle of rotation when △ABC is mapped to △A'B'C'.

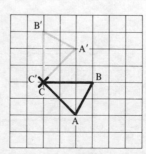

The angle of rotation is 90° anticlockwise.

In questions 1 to 4, give the angle of rotation when △ABC is mapped to △A'B'C'.

1.
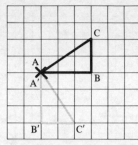

3.
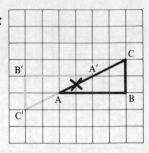

2.
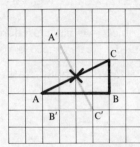

4.
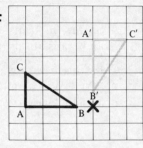

In questions 5 to 10, state the centre of rotation and the angle of rotation. △ABC is the object in each case.

5.
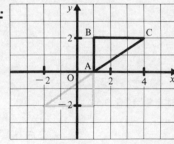

7.
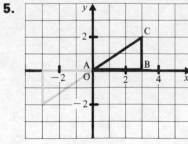

6.
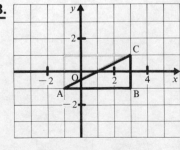

8.
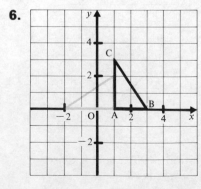

9.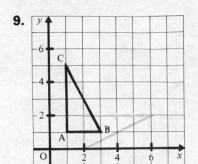

10.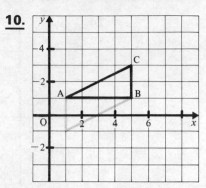

Copy the diagrams in questions 11 to 18, using 1 cm to 1 unit.
Find the images of the given objects under the rotations described.

11.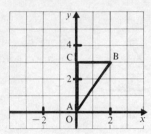

Centre of rotation (0, 0)
Angle of rotation 90° anticlockwise

12.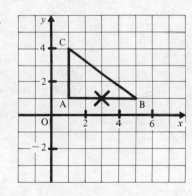

Centre of rotation (3, 1)
Angle of rotation 180°

13.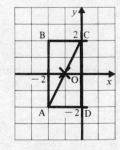

Centre of rotation (−1, 0)
Angle of rotation 180°

14.

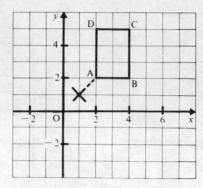

Centre of rotation (1, 1)
Angle of rotation 180°

(As the centre of rotation is not a point on the object, join it to A first.)

15.

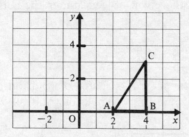

Centre of rotation (0, 0)
Angle of rotation 90° anticlockwise

16.

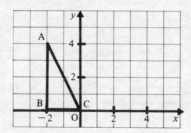

Centre of rotation (2, 0)
Angle of rotation 90° clockwise

17.

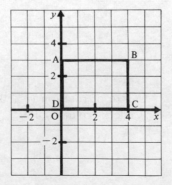

Centre of rotation (2, 0)
Angle of rotation 90° anticlockwise

18.

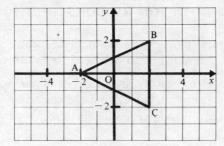

Centre of rotation $(0, 0)$
Angle of rotation $180°$

19. △ABC is rotated about O through $180°$ to give the image, △A'B'C'. Copy and complete the diagram, using 1 cm to 1 unit.

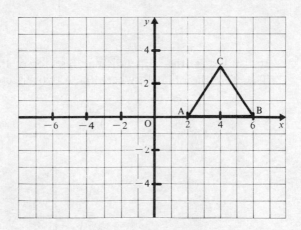

a) What is the shape of the path traced out by C as it moves to C'?

b) Measure OC and OC'. How do they compare?

Repeat with OB and OB'.

20.

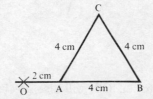

Draw the diagram accurately. Then draw accurately, using a protractor, the image of △ABC under a rotation of $60°$ anticlockwise about O.

FINDING THE CENTRE OF ROTATION BY CONSTRUCTION ⎯⎯⎯⎯⎯

As we have seen we can often spot the centre of rotation just by looking at the diagram but sometimes it is not obvious.

In such cases we can use the fact that an object point A and its image point A' are the same distance from the centre.
So the centre lies on the perpendicular bisector of AA'.

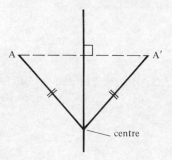

It also lies on the perpendicular bisector of BB'.
Therefore the point P, where these two bisectors meet, is the centre.

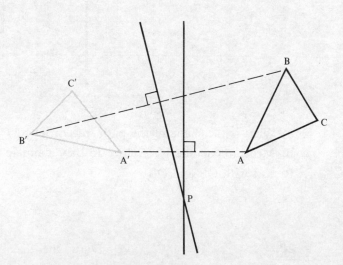

(The perpendicular bisector of CC' will also go through P.)

EXERCISE 9e **1.**

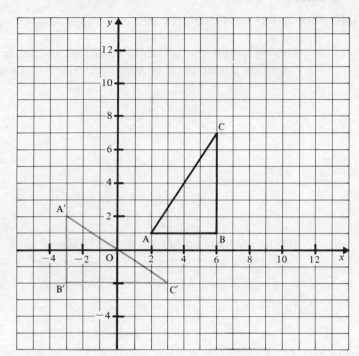

a) Copy the diagram, drawing axes for x and y from -5 to 12. Use 1 cm to 1 unit. $\triangle$A'B'C' is the image of $\triangle$ABC under a rotation.

b) Construct the perpendicular bisectors of AA' and BB'.

c) Mark the centre of rotation, P (that is, the point where the two perpendicular bisectors meet).

d) Check that it is the centre by using tracing paper and the point of your compasses.

e) Join BP and B'P. Measure $B\hat{P}B'$. What is the angle of rotation?

2. Draw axes for x and y from -5 to 10, using 1 cm to 1 unit. Draw $\triangle$ABC with A($-1, 8$), B($5, 4$), C($-1, 1$) and $\triangle$A'B'C' with A'($4, 1$), B'($0, -5$), C'($-3, 1$).

Repeat b) to e) in question 1.

3. Draw axes for x and y from -5 to 10, using 1 cm to 1 unit. Draw $\triangle$ABC with A($-4, -2$), B($2, -2$), C($-4, 4$) and $\triangle$A'B'C' with A'($4, 0$), B'($4, 6$), C'($-2, 0$).

Repeat b) to e) in question 1.

FINDING THE ANGLE OF ROTATION

Having found the centre of rotation, the angle of rotation can be found by joining both an object point and its image to the centre.

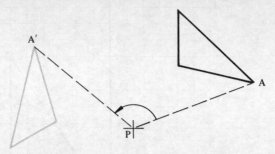

In the diagram above, A′ is the image of A and P is the centre of rotation.

Join both A and A′ to P.   A$\hat{\text{P}}$A′ is the angle of rotation.

In this case the angle of rotation is 120° anticlockwise.

EXERCISE 9f   Trace each of the diagrams and, by drawing in the necessary lines, find the angle of rotation when △ABC is rotated about the centre P to give △A′B′C′.

1.

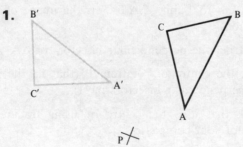

2.

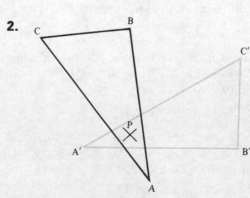

MIXED QUESTIONS ON REFLECTIONS, TRANSLATIONS AND ROTATIONS

EXERCISE 9g

Name the transformation, describing it fully, if the grey triangle is the image of the black one.

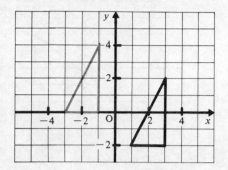

The transformation is a translation given by the vector $\begin{pmatrix} -4 \\ 2 \end{pmatrix}$

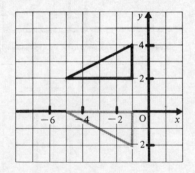

The transformation is a reflection in the line $y = 1$

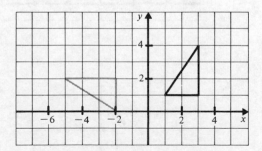

The transformation is a rotation about $(0, -1)$ through an angle of $90°$ anticlockwise

Name the transformations in questions 1 to 10, describing them fully. The black shape is the object, the grey shape is the image.

1.

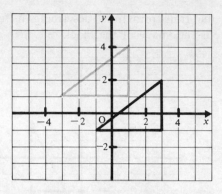

2.

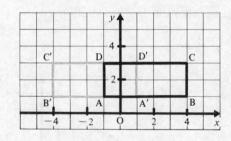

3.

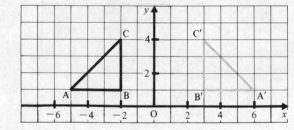

4.

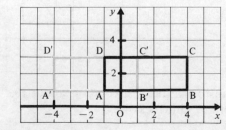

5.

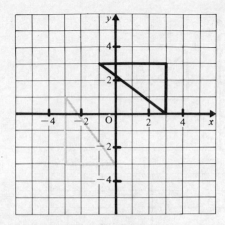

6.

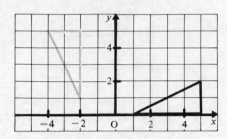

7.

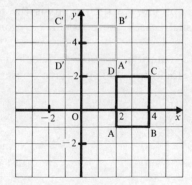

8.

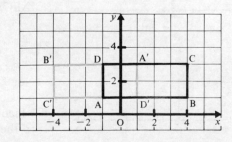

9.

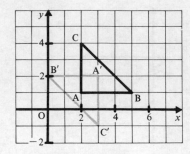

10.

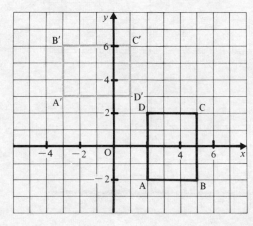

Sometimes we do not know which point is the image of a particular object point. In such cases there could be more than one possible transformation.

(Remember that a rotation of 90° anticlockwise is the same as a rotation of 270° clockwise. Do not give these as two independent transformations.)

11.

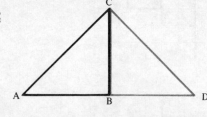

Name and describe two possible transformations which will map the object △ABC to the image △BCD.

12.

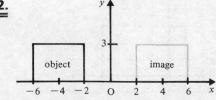

Name and describe three possible transformations which will map the object to the image.

13.

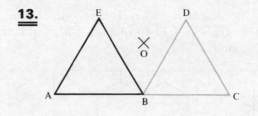

Name and describe four possible transformations which will map the left-hand triangle to the right-hand triangle.

14.

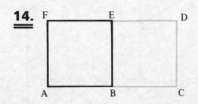

Name and describe five possible transformations which will map the left-hand square to the right-hand square.

15.

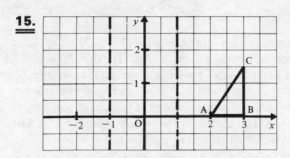

Copy the diagram using 1 cm to 1 unit.
Reflect $\triangle ABC$ in the line $x = 1$ to give $\triangle A_1 B_1 C_1$.
Then reflect $\triangle A_1 B_1 C_1$ in the line $x = -1$ to give $\triangle A_2 B_2 C_2$.
What single transformation will map $\triangle ABC$ to $\triangle A_2 B_2 C_2$?

16. Copy the diagram in question 15 again but draw the axis for x from -5 to 7. Repeat the two reflections but use the line $x = -1$ first and $x = 1$ second. What single transformation is needed this time?

17. A car is turning a corner and two of its positions are shown. Trace the drawing, allowing plenty of space above and below, and find the centre of the turning circle.

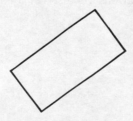
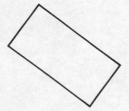

18. Look at the diagram below. Taking one of the shapes as the object, what types of transformations will map it to other shapes in the diagram?

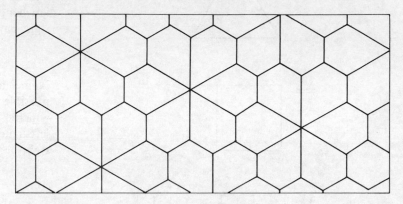

19. Draw axes for x and y from -5 to 5, using $1\,cm$ to 1 unit. Draw lines AB and BC with A(2,2), B(5,2) and C(3,0). Draw the images of ABC under the reflections in the four lines $x = 0$, $x = y$, $y = 0$ and $y = -x$. Draw the images of ABC under the three rotations about O through angles of $90°$, $180°$ and $270°$ anticlockwise. (The seven images of $\triangle$ABC, together with $\triangle$ABC itself, form an eight-pointed star.)

20. Draw axes for x and y from -5 to 5, using $1\,cm$ to 1 unit. Draw $\triangle$ABC with A(2,1), B(4,1) and C(4,2).

a) Reflect $\triangle$ABC in the line $y = x$ to produce the image $\triangle A_1 B_1 C_1$. Then rotate $\triangle A_1 B_1 C_1$ through $180°$ about O to produce $\triangle A_2 B_2 C_2$. What single transformation will map $\triangle$ABC to $\triangle A_2 B_2 C_2$?

b) Rotate $\triangle$ABC through $180°$ about O then reflect the image in the line $y = x$. Is the final image the same as $\triangle A_2 B_2 C_2$?

c) Try other pairs of reflections and rotations, starting a fresh diagram where necessary. In each case find the single transformation which is equivalent to the pair. Does the order in which you do the transformations matter? Are the single transformations themselves all reflections or rotations?

10 AREA

AREA OF A RECTANGLE

Reminder: We can find the area of a rectangle by multiplying its length by its width (or breadth).

$$\text{Area} = \text{length} \times \text{width}$$

or $\quad A = l \times b$

The units we use for the two measurements must be the same.

EXERCISE 10a

Find the area of a rectangle measuring 3.1 cm by 4.2 cm.

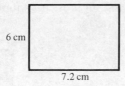

$$\text{Area} = \text{length} \times \text{width}$$
$$= 4.2 \times 3.1 \ \text{cm}^2$$
$$= 13.02 \ \text{cm}^2$$

Find the areas of the rectangles in questions 1 to 16. When finding areas, draw a diagram even if the question is simple.

1.
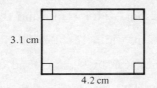

3.
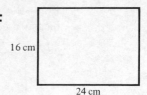

2.
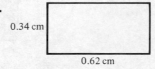

4.
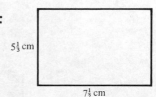

5. Rectangle, length 2.4 cm, width 1.6 cm

6. Rectangle, length 270 cm, breadth 92 cm

7. Rectangle, measuring 0.04 m by 0.02 m

8.   Rectangle, measuring 3.04 m by 1.5 m

9.   Rectangle, measuring $1\frac{1}{2}$ m by $\frac{3}{4}$ m

10. Rectangle, measuring $3\frac{1}{4}$ cm by $1\frac{1}{3}$ cm

Make sure the units are the same before working out the area.

Find the area of a rectangle, measuring 54 mm by 6 cm, in square centimetres.

Width = 54 mm = 5.4 cm

Area = length × width

= 6 × 5.4 cm²

= 32.4 cm²

54 mm

6 cm

11. Rectangle, length 72 mm, width 3 cm. Find the area in cm².

12. Rectangle, length 0.2 m, width 16 cm. Find the area in cm².

13. Rectangle, measuring 0.6 m by 92 mm. Find the area in cm².

14. Rectangle, measuring 420 mm by 16 cm. Find the area in cm².

15. Rectangle, measuring 41 mm by 7 cm. Find the area in mm².

16. Rectangle, measuring 1246 cm by 69.2 m. Find the area in m².

Find the areas of the following figures in square centimetres. The measurements are all in centimetres.

17.
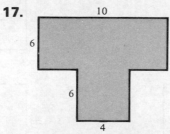

19.
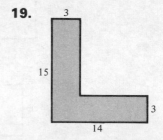

18.
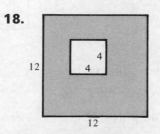

20.
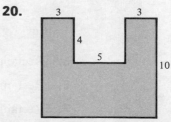

FINDING A LENGTH WHEN THE AREA IS GIVEN

EXERCISE 10b

Find the length of a rectangle of area 20 cm^2 and width 2.5 cm.

20 cm²   2.5 cm

Area $= \text{length} \times \text{width}$, or $A = l \times b$,

then $\text{length} = \dfrac{\text{area}}{\text{width}}$ or $l = \dfrac{A}{b}$

Length $= \dfrac{20}{2.5}$ cm

$= \dfrac{200}{25}$ cm

$= 8 \text{ cm}$

Find the missing measurements for the following rectangles:

	Area	Length	Width
1.	2.4 cm^2	6 cm	
2.	20 cm^2	4 cm	
3.	36 m^2		3.6 m
4.	108 mm^2	27 mm	
5.	3 cm^2		0.6 cm
6.	6 m^2	4 m	
7.	20 cm^2		16 cm
8.	7.2 m^2		2.4 m
9.	4.2 m^2		0.6 m
10.	14.4 cm^2	2.4 cm	

AREA OF A PARALLELOGRAM

Knowing how to find the area of a rectangle helps us to deal with parallelograms.

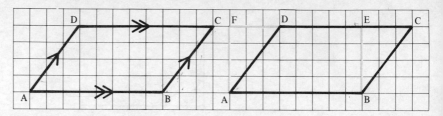

Copy the first diagram on to squared paper. Draw the line BE and remove △BEC from the right-hand side. Draw an equal triangle, FDA, at the left-hand side to replace △BEC. Then you can see that the area of the parallelogram ABCD is equal to the area of rectangle ABEF.

Area of parallelogram = AB × BE

= base × perpendicular height

When we use the word *height* we mean the perpendicular height BE, not the slant height BC, so we can say

Area of parallelogram = base × height

EXERCISE 10c

Find the area of a parallelogram of base 7 cm, height 5 cm and slant height 6 cm.

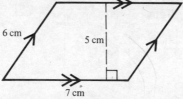

Area = base × height

= 7 × 5 cm²

= 35 cm²

(Notice that we do not need the length of the 6 cm side.)

Find the areas of the following parallelograms:

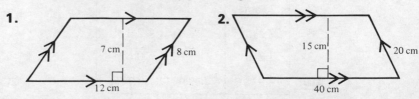

1.

2.

3.

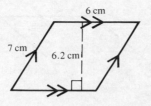

7 cm
6 cm
6.2 cm

6.

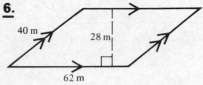

40 m
28 m
62 m

4.

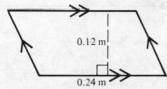

0.12 m
0.24 m

7.

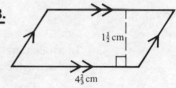

4.3 cm
3.4 cm
7.2 cm

5.

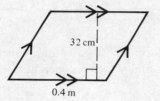

32 cm
0.4 m

8.

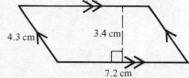

$1\frac{1}{2}$ cm
$4\frac{2}{3}$ cm

Find the area of the parallelogram.

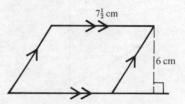

$7\frac{1}{2}$ cm
6 cm

(Notice that it does not matter where the height is measured.)

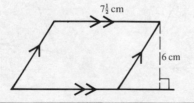

$7\frac{1}{2}$ cm
6 cm

Area = base × height

$= 7\frac{1}{2} \times 6 \,\text{cm}^2$

$= \frac{15}{2} \times \overset{3}{6}\,\text{cm}^2$

$= 45\,\text{cm}^2$

9.

6 cm
3.6 cm
10.8 cm

10.

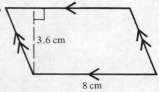

3.6 cm
8 cm

11.

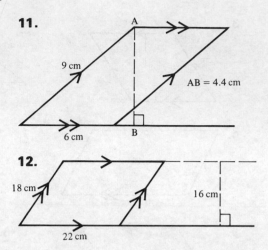

9 cm

AB = 4.4 cm

6 cm

B

12.

18 cm

16 cm

22 cm

In questions 13 to 18, turn the page round if necessary so that you can see which is the base and which the height.

13.

9 cm

7 cm

8 cm

16.

13 cm

10 cm

12 cm

14.

6 cm

4 cm

12 cm

17.

7.2 cm

5 cm

7 cm

15.

2 cm

2.5 cm

4.5 cm

18.

12 cm

9 cm

20 cm

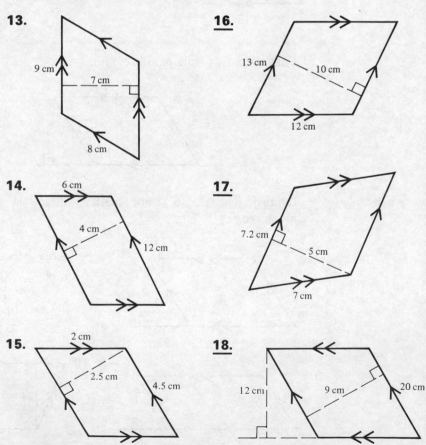

For each of the following questions, draw axes for x and y for values from -4 to 4. Use 1 square to 1 unit. Draw parallelogram ABCD and find its area in square units.

19.   A$(-2, 0)$, B$(2, 0)$, C$(3, 2)$, D$(-1, 2)$

20.   A$(2, -2)$, B$(4, 1)$, C$(-1, 1)$, D$(-3, -2)$

21.   A$(2, 1)$, B$(2, 4)$, C$(-1, 2)$, D$(-1, -1)$

22.   A$(2, 0)$, B$(2, 3)$, C$(-3, 4)$, D$(-3, 1)$

THE AREA OF A TRIANGLE

There are two ways of finding how to calculate the area of a triangle.

First, if we think of a triangle as half a parallelogram we get

area of triangle $= \frac{1}{2} \times$ area of parallelogram

$\qquad\qquad\quad = \frac{1}{2}$(base $\times$ height)

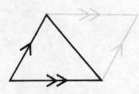

Second, if we enclose the triangle in a rectangle we see again that the area of the triangle is half the area of the rectangle.

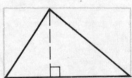

These diagrams can be drawn on squared paper and then cut out to show how the pieces fit.

THE HEIGHT OF A TRIANGLE

As with the parallelogram, when we talk about the height of a triangle we mean its perpendicular height and not its slant height.

If we draw the given triangle accurately on squared paper, we can see that the height of the triangle is not 10 cm or 7.5 cm but 6 cm. (We can also see that the foot of the perpendicular is *not* the midpoint of the base.)

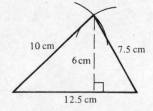

FINDING AREAS OF TRIANGLES

EXERCISE 10d

Find the area of a triangle with base 7 cm and height 6 cm.

Area $= \frac{1}{2}$ (base × height)

$= \frac{1}{2} \times 7 \times 6 \text{ cm}^2$

$= 21 \text{ cm}^2$

Find the areas of the following triangles.

1.
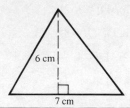
8 cm
12 cm

5.
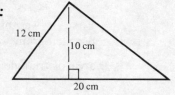
12 cm
10 cm
20 cm

2.
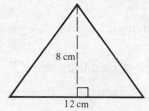
1.2 m
2.6 m

6.
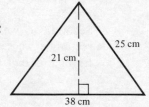
25 cm
21 cm
38 cm

3.
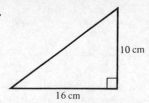
10 cm
16 cm

7.
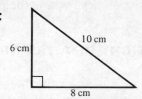
10 cm
6 cm
8 cm

4.
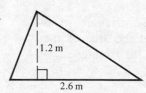
2.4 cm
2 cm
3.2 cm

8.
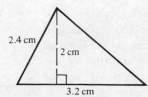
4.8 cm
6 cm

Find the area of the given triangle.

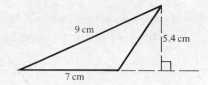

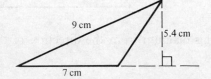

Area = $\frac{1}{2}$ (base × height)

= $\frac{1}{2}$ × 7 × 5.4 cm²

= 18.9 cm²

9.

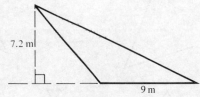

10.

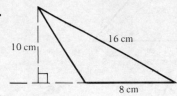

11.

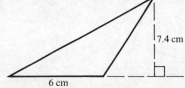

12.

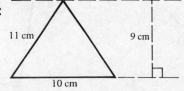

If necessary turn the page round and look at the triangle from a different direction.

Find the area of the triangle.

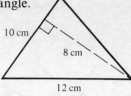

(Look at this diagram from the direction of the arrow.)

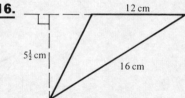

Area = $\frac{1}{2}$ (base × height)
= $\frac{1}{2} \times 10 \times 8$ cm²
= 40 cm²

13.

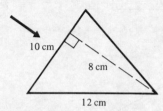

16.

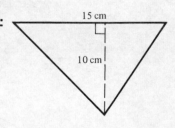

14.

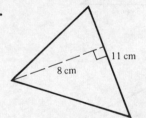

17.

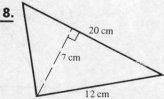

15.

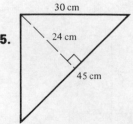

18.

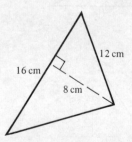

19.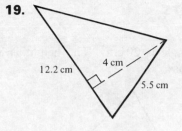

22.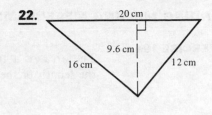

20.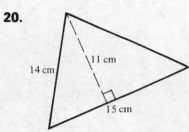

23.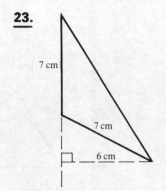

21.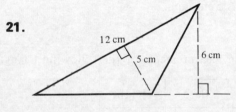

24.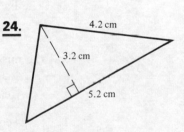

For questions 25 to 30, use squared paper to draw axes for *x* and *y* from 0 to 6 using 1 square to 1 unit. Find the area of each triangle.

25. △ABC with A(1, 0), B(6, 0) and C(4, 4)

26. △PQR with P(0, 2), Q(6, 0) and R(6, 4)

27. △DEF with D(1, 1), E(1, 5) and F(6, 0)

28. △LMN with L(5, 0), M(0, 6) and N(5, 6)

29. △ABC with A(0, 5), B(5, 5) and C(4, 1)

30. △PQR with P(2, 1), Q(2, 6) and R(5, 3)

FINDING MISSING MEASUREMENTS

EXERCISE 10e

The area of a triangle is $20\,\text{cm}^2$. The height is $8\,\text{cm}$. Find the length of the base.

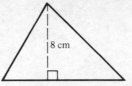

Let the base be $b\,\text{cm}$ long.

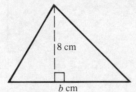

$\text{Area} = \frac{1}{2}\,(\text{base} \times \text{height})$

$20 = \frac{1}{2} \times b \times 8^{4}$

$20 = 4b$

$b = 5$

The base is $5\,\text{cm}$ long.

Find the missing measurements of the following triangles.

	Area	Base	Height
1.	$24\,\text{cm}^2$	$6\,\text{cm}$	
2.	$30\,\text{cm}^2$		$10\,\text{cm}$
3.	$48\,\text{cm}^2$		$16\,\text{cm}$
4.	$10\,\text{cm}^2$	$10\,\text{cm}$	
5.	$36\,\text{cm}^2$	$24\,\text{cm}$	
6.	$108\,\text{cm}^2$		$6\,\text{cm}$
7.	$96\,\text{cm}^2$		$64\,\text{cm}$
8.	$4\,\text{cm}^2$		$3\,\text{cm}$
9.	$2\,\text{cm}^2$	$10\,\text{cm}$	
10.	$1.2\,\text{cm}^2$	$0.4\,\text{cm}$	
11.	$72\,\text{cm}^2$		$18\,\text{cm}$
12.	$1.28\,\text{cm}^2$	$0.64\,\text{cm}$	

COMPOUND SHAPES

EXERCISE 10f

ABCE is a square of side 8 cm. The total height of the shape is 12 cm. Find the area of ABCDE.

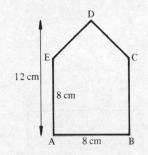

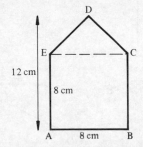

The height of the triangle is 4 cm.

Area of $\triangle ECD = \frac{1}{2}$ (base $\times$ height)

$$= \frac{1}{\cancel{2}_1} \times \cancel{8} \times \overset{4}{4} \, \text{cm}^2$$

$$= 16 \, \text{cm}^2$$

Area of ABCE $= 8 \times 8 \, \text{cm}^2$

$$= 64 \, \text{cm}^2$$

Total area $= 80 \, \text{cm}^2$

Find the areas of the following shapes. Remember to draw a diagram for each question and mark in all the measurements.

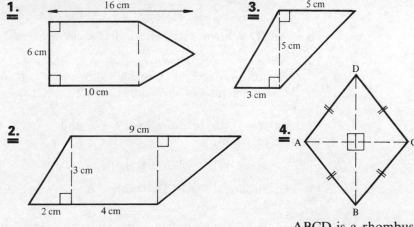

ABCD is a rhombus.
AC = 9 cm.
BD = 12 cm.

5.

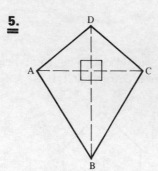

ABCD is a kite.
(BD is the axis of symmetry.
The diagonals cut at right angles.)
AC = 10 cm.   BD = 12 cm.

6.

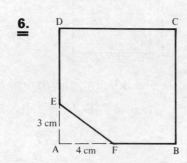

A square ABCD, of side 9 cm, has
a triangle EAF cut off it.

7.

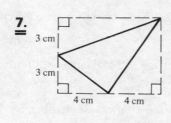

8.

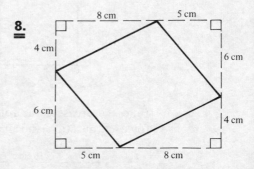

9. ABCD is a rhombus whose diagonals measure 7 cm and 11 cm.

10. ABCD is a kite whose diagonals measure 12 cm and 8 cm.
(There are several possible kites you can draw with these
measurements but their areas are all the same.)

In questions 11 to 16 draw axes for x and y from -6 to $+6$, using
1 square to 1 unit.

Find the areas of the following shapes:

11. Quadrilateral   ABCD   with   A$(-2, -3)$,   B$(3, -3)$,   C$(0, 4)$
and D$(-2, 4)$

12. Quadrilateral EFGH with E$(-1, 1)$, F$(2, -3)$, G$(5, 1)$, and
H$(2, 5)$

13. Pentagon IJKLM with I(−1, −4), J(2, −4), K(4, −1), L(2, 1) and M(−1, 1)

14. Quadrilateral PQRS with P(−1, 2), Q(1, −3), R(3, 2), and S(1, 4)

15. Pentagon TUVWZ with T(−2, 0), U(0, −2), V(4, −2), W(4, 3) and Z(−2, 3)

16. Triangle ABC with A(3, 0), B(4, 3) and C(−3, 2)

MIXED EXERCISES

EXERCISE 10g Find the areas of the following figures:

1.

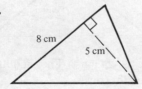

3.

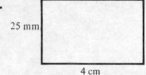

2.

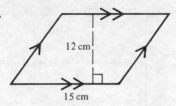

4.

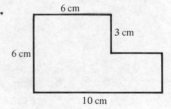

5. The area of a rectangle is 84 cm² and its width is 6 cm. Find its length.

6. The area of the parallelogram below is 52 cm². Find the distance, *d* cm, between the parallel lines.

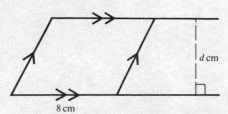

EXERCISE 10h Find the areas of the following figures:

1.

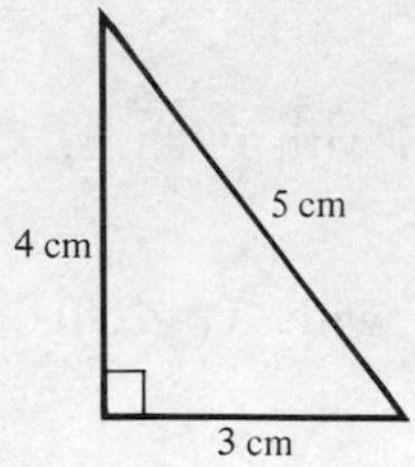

3.

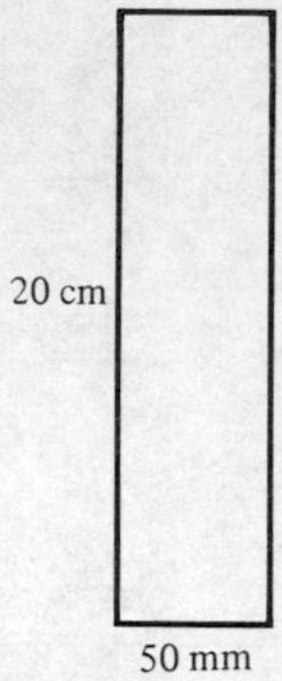

2.

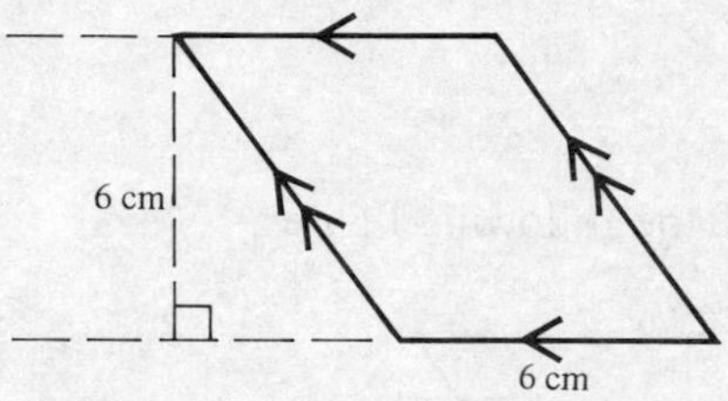

4.

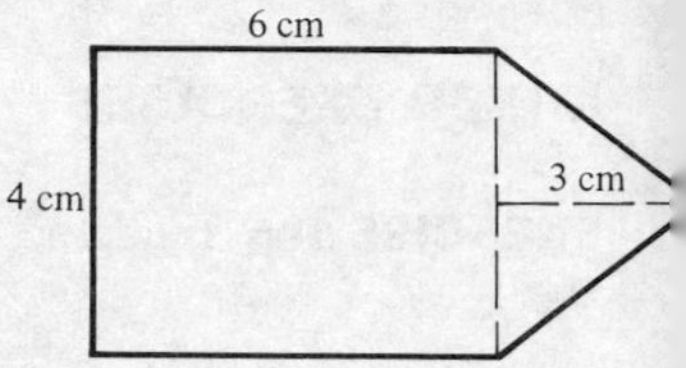

5.

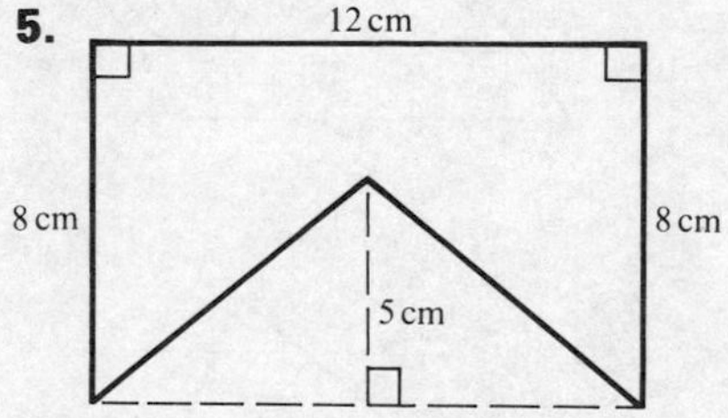

6. The area of a triangle is 24 cm^2. The height of the triangle is 8 cm. Find the length of the base.

11 CIRCLES CIRCUMFERENCE AND AREA

DIAMETER, RADIUS AND CIRCUMFERENCE

When you use a pair of compasses to draw a circle, the place where you put the point is the *centre* of the circle. The line that the pencil draws is the *circumference* of the circle.

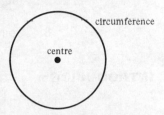

Any straight line joining the centre to a point on the circumference is a *radius*.

A straight line across the full width of a circle (i.e. going through the centre) is a *diameter*.

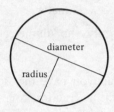

The diameter is twice as long as the radius. If d stands for the length of a diameter and r stands for the length of a radius, we can write this as a formula:

$$d = 2r$$

EXERCISE 11a In questions 1 to 5, write down the length of the diameter of the circle whose radius is given

1.

6 cm

2.

5 m

3. 15 mm

4. 3.5 cm

5. 1 km

6. 4.6 cm

7. For this question you will need some thread and a cylinder (e.g. a tin of soup, a soft drink can, the cardboard tube from a roll of kitchen paper).
Measure across the top of the cylinder to get a value for the diameter. Wind the thread 10 times round the can. Measure the length of thread needed to do this and then divide your answer

by 10 to get a value for the circumference. If C stands for the length of the circumference and d for the length of the diameter, find, approximately, the value of $C \div d$.

(Note that you can also use the label from a cylindrical tin. If you are careful you can reshape it and measure the diameter and then unroll it to measure the circumference.)

8.   Compare the results from the whole class for the value of $C \div d$.

INTRODUCING π

From the last exercise you will see that, for any circle,

$$\text{circumference} \approx 3 \times \text{diameter}$$

The number that you have to multiply the diameter by to get the circumference is slightly larger than 3.

This number is unlike any number that you have met so far. It cannot be written down exactly, either as a fraction or as a decimal: as a fraction it is approximately, but *not* exactly, $\frac{22}{7}$; as a decimal it is approximately 3.142, which is correct to 3 decimal places.

Over the centuries mathematicians have spent a lot of time trying to find the true value of this number. The ancient Chinese used 3. Three is also the value given in the Old Testament (1 Kings 7:23). The Egyptians ($c.$ 1600 BC) used $4 \times (\frac{8}{9})^2$. Archimedes ($c.$ 225 BC) was the first person to use a sound method for finding its value and a mathematician called Van Ceulen (1540–1610) spent most of his life finding it to 35 decimal places!

Now with a computer to do the arithmetic we can find its value to as many decimal places as we choose: it is a never ending, never repeating decimal fraction. To as many figures as we can get across the page, the value of this number is

3.141592653589793238462643383279502884197169399375105820974944

Because we cannot write it down exactly we use the Greek letter π (pi) to stand for this number. Then we can write a formula connecting the circumference and diameter of a circle in the form $C = \pi d$. But $d = 2r$ so we can rewrite this formula as

$$C = 2\pi r$$

where $C = $ circumference and $r = $ radius

CALCULATING THE CIRCUMFERENCE

EXERCISE 11b

Using 3.142 as an approximate value for π, find the circumference of a circle of radius 3.8 m.

3.8 m

Using $C = 2\pi r$

with $\pi = 3.142$ and $r = 3.8$

gives $C = 2 \times 3.142 \times 3.8$

$= 23.9$ to 3 s.f.

Circumference $= 23.9$ m to 3 s.f.

Using 3.142 as an approximate value for π and giving your answers correct to 3 s.f., find the circumference of a circle of radius:

1.	2.3 m	**6.**	250 mm	**11.**	7 cm
2.	4.6 cm	**7.**	36 cm	**12.**	28 mm
3.	2.9 cm	**8.**	4.8 m	**13.**	1.4 m
4.	53 mm	**9.**	1.8 m	**14.**	35 mm
5.	8.7 m	**10.**	0.014 km	**15.**	5.6 cm

Find the circumference of a circle of diameter 12.6 mm. Take $\pi \approx 3.142$.

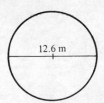

12.6 m

Method 1:

Using $C = 2\pi r$,

$r = \frac{1}{2}$ of $12.6 = 6.3$

$C = 2 \times 3.142 \times 6.3$

$= 39.6$ to 3 s.f.

Method 2:

Using $C = \pi d$

$C = 3.142 \times 12.6$

$= 39.6$ to 3 s.f.

Circumference $= 39.6$ mm to 3 s.f.

Using $\pi \approx 3.14$ and giving your answer correct to 2 s.f., find the circumference of a circle of:

16. radius 154 mm

20. radius 34.6 cm

17. diameter 28 cm

21. diameter 511 mm

18. diameter 7.7 m

22. diameter 630 cm

19. radius 210 mm

23. diameter 9.1 m

PROBLEMS

EXERCISE 11c Use 3.142 as an approximate value for π and give your answers correct to 3 s.f.

Find the perimeter of the given
semicircle.
(The prefix "semi" means half.)

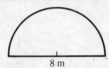

8 m

The complete circumference of the circle is $2\pi r$

The curved part of the semicircle is $\frac{1}{2} \times 2\pi r$

$$= \frac{1}{2} \times 2 \times 3.142 \times 4 \, \text{m}$$

$$= 12.56 \, \text{m}$$

The perimeter = curved part + straight edge

$$= (12.56 + 8) \, \text{m}$$

$$= 20.56 \, \text{m}$$

$$= 20.6 \, \text{m} \quad \text{to 3 s.f.}$$

Find the perimeter of each of the following shapes:

1.   4 cm

2.

3 cm

(This is called a *quadrant*:
it is one quarter of a circle.)

3.

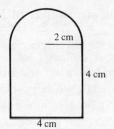

2 cm

4 cm

4 cm

7.

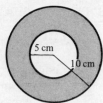

5 cm

10 cm

4.

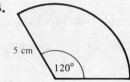

5 cm

120°

(This is one third of a circle because 120° is $\frac{1}{3}$ of 360°.)

8.

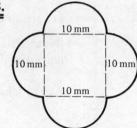

10 mm

10 mm

10 mm

10 mm

5.

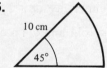

10 cm

45°

A "slice" of a circle is called a sector.
($\frac{45}{360} = \frac{1}{8}$, so this sector is $\frac{1}{8}$ of a circle.)

9.

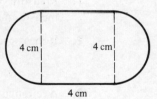

4 cm

4 cm

4 cm

6.

5 cm

10.

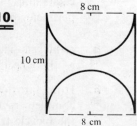

8 cm

10 cm

8 cm

EXERCISE 11d Take π as 3.142 and give your answers correct to 3 s.f.

A circular flower bed has a diameter of 1.5 m. A metal edging is to be placed round it. Find the length of edging needed and the cost of the edging if it is sold by the metre (i.e. you can only buy a whole number of metres) and costs 60 p a metre.

Using $C = \pi d$,

$$C = 3.142 \times 1.5$$

$$= 4.71$$

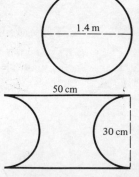

Length of edging needed = 4.71 m   to 3 s.f.

(Note that if you use $C = 2\pi r$, you must remember to halve the diameter.)

As the length is 4.71 m we have to buy 5 m of edging.

$$\text{Cost} = 5 \times 60 \, \text{p}$$

$$= 300 \, \text{p} \quad \text{or} \quad £3$$

---

1. Measure the diameter, in millimetres, of a 2 p coin. Use your measurement to find the circumference of a 2 p coin.

2. Repeat question 1 with a 10 p coin and a 1 p coin.

3. A circular table cloth has a diameter of 1.4 m. How long is the hem of the cloth?

4. A rectangular sheet of metal measuring 50 cm by 30 cm has a semicircle of radius 15 cm cut from each short side as shown. Find the perimeter of the shape that is left.

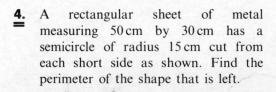

5. A bicycle wheel has a radius of 28 cm. What is the circumference of the wheel?

6. How far does a bicycle wheel of radius 28 cm travel in one complete revolution? How many times will the wheel turn when the bicycle travels a distance of 352 m?

7. A cylindrical tin has a radius of 2 cm. What length of paper is needed to put a label on the tin if the edges just meet?

8. A square sheet of metal has sides of length 30 cm. A quadrant (one quarter of a circle) of radius 15 cm is cut from each of the four corners. Sketch the shape that is left and find its perimeter.

9. A boy flies a model aeroplane on the end of a wire 10 m long. If he keeps the wire horizontal, how far does his aeroplane fly in one revolution?

10. If the aeroplane described in question 9 takes 1 second to fly 10 m, how long does it take to make one complete revolution? If the aeroplane has enough power to fly for 1 minute, how many turns can it make?

11. A cotton reel has a diameter of 2 cm. There are 500 turns of thread on the reel. How long is the thread?

12. A bucket is lowered into a well by unwinding rope from a cylindrical drum. The drum has a radius of 20 cm and with the bucket out of the well there are 10 complete turns of the rope on the drum. When the rope is fully unwound the bucket is at the bottom of the well. How deep is the well?

13. A garden hose is 100 m long. For storage it is wound on a circular hose reel of diameter 45 cm. How many turns of the reel are needed to wind up the hose?

14. The cage which takes miners up and down the shaft of a coal mine is raised and lowered by a rope wound round a circular drum of diameter 3 m. It takes 10 revolutions of the drum to lower the cage from ground level to the bottom of the shaft. How deep is the shaft?

FINDING THE RADIUS OF A CIRCLE GIVEN THE CIRCUMFERENCE ▬

If a circle has a circumference of 24 cm, we can find its radius from the formula $C = 2\pi r$ either by using the formula as it stands,

i.e. $24 = 2 \times 3.142 \times r$

and solving this equation for r

or by first making r the subject of $C = 2\pi r$ as follows

Divide both sides by 2 and π $C = 2 \times \pi \times r$

$$\frac{C}{1} \times \frac{1}{2 \times \pi} = \frac{{}^{1}\cancel{2}}{1} \times \frac{\cancel{\pi}^{1}}{1} \times \frac{r}{1} \times \frac{1}{\cancel{2} \times \cancel{\pi}}$$

$$\frac{C}{2\pi} = r$$

i.e.

$$\boxed{r = \frac{C}{2\pi}}$$

EXERCISE 11e Take π as 3.142 and give your answers correct to 3 s.f.

The circumference of a circle is 36 cm. Find the radius of this circle.

Either: Using $C = 2\pi r$ gives

$$36 = 2 \times 3.142 \times r$$

$$36 = 6.284 \times r$$

$$\frac{36}{6.284} = r \quad \text{(dividing both sides by 6.284)}$$

$$r = 5.73 \quad \text{to 3 s.f.}$$

Or: Using $r = \dfrac{C}{2\pi}$ gives

$$r = \frac{\cancel{36}^{18}}{\cancel{2} \times 3.142}$$

$$= 5.73 \quad \text{to 3 s.f.}$$

Therefore the radius is 5.73 cm correct to 3 s.f.

Find the radius of the circle whose circumference is:

1. 44 cm

2. 121 mm

3. 550 m

4. 275 cm

5. 462 mm

6. 831 cm

7. 36.2 mm

8. 391 m

9. 582 cm

10. 87.4 m

11. Find the diameter of the circle whose circumference is 52 m.

12. A roundabout at a major road junction is to be built. It has to have a minimum circumference of 188 m. What is the corresponding minimum diameter?

13. A bicycle wheel has a circumference of 200 cm. What is the radius of the wheel?

14. A car has a turning circle whose circumference is 63 m. What is the narrowest road that the car can turn round in without going on the pavement?

15. When the label is taken off a tin of soup it is found to be 32 cm long. If there was an overlap of 1 cm when the label was on the tin, what is the radius of the tin?

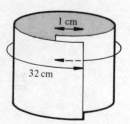

16. The diagram shows a quadrant of a circle. If the curved edge is 15 cm long, what is the length of a straight edge?

17. A tea cup has a circumference of 24 cm. What is the radius of the cup? Six of these cups are stored edge to edge in a straight line on a shelf. What length of shelf do they occupy?

18. Make a cone from a sector of a circle as follows:

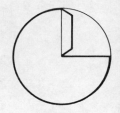

On a sheet of paper draw a circle of radius 8 cm. Draw two radii at an angle of 90°. Make a tab on one radius as shown. Cut out the larger sector and stick the straight edges together. What is the circumference of the circle at the bottom of the cone?

19. A cone is made by sticking together the straight edges of the sector of a circle, as shown in the diagram.

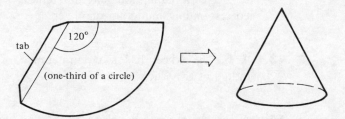

The circumference of the circle at the bottom of the finished cone is 10 cm. What is the radius of the circle from which the sector was cut?

20. The shape in the diagram is made up of a semicircle and a square. Find the length of a side of this square.

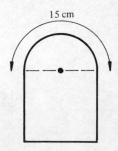

21. The curved edge of a sector of angle 60° is 10 cm. Find the radius and the perimeter of the sector.

THE AREA OF A CIRCLE

The formula for finding the area of a circle is

$$A = \pi r^2$$

You can see this if you cut a circle up into sectors and place the pieces together as shown to get a shape which is roughly rectangular. Consider a circle of radius r whose circumference is $2\pi r$.

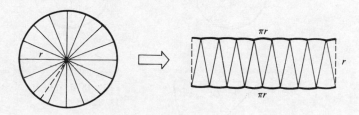

Area of circle = area of "rectangle"

$= \text{length} \times \text{width}$

$= \pi r \times r = \pi r^2$

EXERCISE 11f Take π as 3.142 and give your answers correct to 3 s.f.

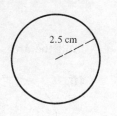

Find the area of a circle of radius 2.5 cm.

Using $A = \pi r^2$

with $\pi = 3.142$ and $r = 2.5$

gives $A = 3.142 \times (2.5)^2$

$= 19.6$ to 3 s.f.

Area is 19.6 cm^2 to 3 s.f.

Find the areas of the following circles:

1.

4 cm

2.

8 cm

3.

5 m

4.

10 mm

(be careful!)

6.

60 cm

8.

3.5 km

5.

7 cm

7.

3.8 m

9.

80 m

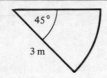

45°

3 m

This is a *sector* of a circle. Find its area.

$$\frac{\overset{9}{\cancel{45}}}{\underset{72}{\cancel{360}}} = \frac{\overset{1}{\cancel{9}}}{\underset{8}{\cancel{72}}} = \frac{1}{8}$$

∴ area of sector $= \frac{1}{8}$ of area of circle of radius 3 m

Area of sector $= \frac{1}{8}$ of πr^2

$= \frac{1}{8} \times 3.142 \times 9 \, \text{m}^2$

$= 3.53 \, \text{m}^2$ to 3 s.f.

Find the areas of the following shapes:

10.

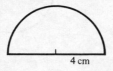

4 cm

12.

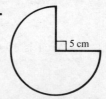

5 cm

11.

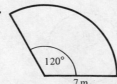

120°

7 m

13.

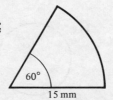

60°

15 mm

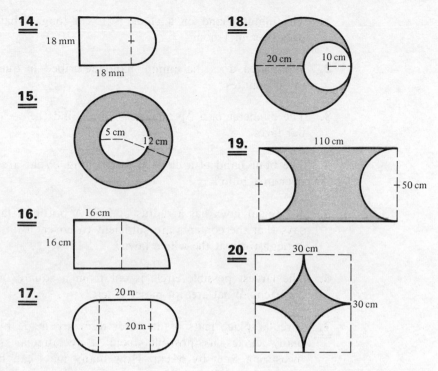

14.

18 mm

18 mm

15.

5 cm

12 cm

16. 16 cm

16 cm

17. 20 m

20 m

18.

20 cm 10 cm

19. 110 cm

50 cm

20. 30 cm

30 cm

PROBLEMS

EXERCISE 11g Take π as 3.142 and make a rough sketch to illustrate each problem. Give your answers to 3 s.f.

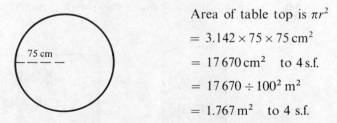

A circular table has a radius of 75 cm. Find the area of the table top. The top of the table is to be varnished. One tin of varnish covers $4\,\text{m}^2$. Will one tin be enough to give the table top three coats of varnish?

75 cm

Area of table top is πr^2

$= 3.142 \times 75 \times 75\,\text{cm}^2$

$= 17\,670\,\text{cm}^2$ to 4 s.f.

$= 17\,670 \div 100^2\,\text{m}^2$

$= 1.767\,\text{m}^2$ to 4 s.f.

For three coats, enough varnish is needed to cover
$3 \times 1.767\,\text{m}^2 = 5.30\,\text{m}^2$ to 3 s.f.

So one tin of varnish is not enough.

1. The minute hand on a clock is 15 cm long. What area does it pass over in 1 hour?

2. What area does the minute hand described in question 1 cover in 20 minutes?

3. The diameter of a 2 p coin is 25 mm. Find the area of one of its flat faces.

4. The hour hand of a clock is 10 cm long. What area does it pass over in 1 hour?

5. A circular lawn has a radius of 5 m. A bottle of lawn weedkiller says that the contents are sufficient to cover 50 m². Is one bottle enough to treat the whole lawn?

6. The largest possible circle is cut from a square of paper 10 cm by 10 cm. What area of paper is left?

7. Circular place mats of diameter 8 cm are made by stamping as many circles as possible from a rectangular strip of card measuring 8 cm by 64 cm. How many mats can be made from the strip of card and what area of card is wasted?

8. A wooden counter top is a rectangle measuring 280 cm by 45 cm. There are three circular holes in the counter, each of radius 10 cm. Find the area of the wooden top.

9. The surface of the counter top described in question 8 is to be given four coats of varnish. If one tin of varnish covers 3.5 m², how many tins will be needed?

10. Take a cylindrical tin of food with a paper label:

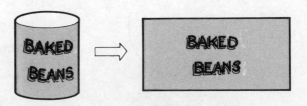

Measure the diameter of the tin and use it to find the length of the label. Find the area of the label. Now find the total surface area of the tin (two circular ends and the curved surface).

11.   *Count Buffon's experiment*

Count Buffon was an eighteenth-century scientist who carried out many probability experiments. The most famous of these is his "Needle Problem". He dropped needles on to a surface ruled with parallel lines and considered the drop successful when a needle fell across a line and unsuccessful when a needle fell between two lines. His amazing discovery was that the number of successful drops divided by the number of unsuccessful drops was an expression involving π.

You can repeat his experiment and get a good approximation for the value of π from it:

Take a matchstick or a similar small stick and measure its length. Now take a sheet of paper measuring about $\frac{1}{2}$ m each way and fill the sheet with a set of parallel lines whose distance apart is equal to the length of the stick. With the sheet on the floor drop the stick on to it from a height of about 1 m. Repeat this about a hundred times and keep a tally of the number of times the stick touches or crosses a line and of the number of times it is dropped. Then find the value of

$$\frac{2 \times \text{number of times it is dropped}}{\text{number of times it crosses or touches a line}}$$

MIXED EXERCISES

Take π as 3.142. Give your answers to 3 s.f.

EXERCISE 11h   **1.**   Find the circumference of a circle of radius 2.8 mm.

2.   Find the radius of a circle of circumference 60 m.

3.   Find the circumference of a circle of diameter 12 cm.

4.   Find the area of a circle of radius 2.9 m.

5.   Find the area of a circle of diameter 25 cm.

6.

8 mm

Find the perimeter of the quadrant in the diagram.

7.

45°

4.5 cm

Find the area of the sector in the diagram.

EXERCISE 11i   **1.** Find the circumference of a circle of diameter 20 m.

2. Find the area of a circle of radius 12 cm.

3. Find the radius of a circle of circumference 360 cm.

4. Find the area of a circle of diameter 8 m.

5. Find the diameter of a circle of circumference 280 mm.

6. Find the perimeter of the sector in the diagram.

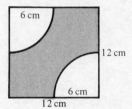

7. Find the area of the shaded part of the diagram.

EXERCISE 11j   **1.** Find the area of a circle of radius 2 km.

2. Find the circumference of a circle of radius 49 mm.

3. Find the radius of a circle of circumference 88 m.

4. Find the area of a circle of diameter 14 cm.

5. Find the area of a circle of radius 3.2 cm.

6.

An ornamental pond in a garden is a rectangle with a semicircle on each short end. The rectangle measures 5 m by 3 m and the radius of each semicircle is 1 m. Find the area of the pond.

7. A formal flower bed is square in shape with a quarter circle cut from each corner. The square (before the quadrants are removed) measures 4 m by 4 m. The radius of each quadrant is 1 m. Sketch the flower bed and find its perimeter.

12 RATIO

SIMPLIFYING RATIOS

Suppose that Peter makes a model of his father's boat. If the model is 1 m long while the actual boat is 20 m long, we say that the ratio of the length of the model to the length of the actual boat is 1 m : 20 m or, more simply, 1 : 20. We can also write the ratio as the fraction $\frac{1}{20}$.

If Peter built a larger model which was 2 m long then the ratio would be

$$\frac{\text{length of model}}{\text{length of actual boat}} = \frac{2\,\text{m}}{20\,\text{m}} = \frac{1}{10}$$

or   length of model : length of boat = 1 : 10

Ratios are therefore comparisons between related quantities.

EXERCISE 12a

Express the ratios   a) 24 to 72   b) 2 cm to 1 m   in their simplest form.

a) $\dfrac{24}{72} = \dfrac{3}{9} = \dfrac{1}{3}$ (dividing both numbers by 8 and then by 3)

  or   24 : 72 = 3 : 9 = 1 : 3
  so   24 : 72 = 1 : 3

b) (Before we can compare 2 cm and 1 m they must be expressed in the same unit.)

$\dfrac{2\,\text{cm}}{1\,\text{m}} = \dfrac{2\,\text{cm}}{100\,\text{cm}}$ or   2 cm : 1 m = 2 cm : 100 cm

$\qquad\qquad = \dfrac{1}{50}$ $\qquad\qquad\qquad\qquad = 2 : 100$

$\qquad\qquad\qquad\qquad\qquad\qquad\qquad\qquad = 1 : 50$

  so   2 cm : 1 m = 1 : 50

Express the following ratios in their simplest form:

1. 8 : 10

2. 20 : 16

3. 12 : 18

4. 2 cm : 8 cm

5. 32 p : 96 p

6. 45 g : 1 kg

7. £4 : 75 p

9. 288 : 306

8. 48 p : £2.88

10. $10 \, cm^2 : 1 \, m^2$

Simplify the ratio 24 : 18 : 12

(As there are three numbers involved, this ratio cannot be expressed as a single fraction.)

24 : 18 : 12 = 4 : 3 : 2   (dividing each number by 6)

---

11. 4 : 6 : 10

16. 7 : 56 : 49

12. 18 : 24 : 36

17. 15 : 20 : 35

13. 2 : 10 : 20

18. 16 : 128 : 64

14. 9 : 12 : 15

19. 144 : 12 : 24

15. 20 : 24 : 32

20. 98 : 63 : 14

We know that we can produce equivalent fractions by multiplying or dividing both numerator and denominator by the same number,

so that $\frac{2}{3} = \frac{4}{6}$ or $\frac{12}{18}$ or $\frac{20}{30}$.

We can do the same with a ratio in the form 3 : 6.

3 : 6 = 6 : 12   (multiplying both numbers by 2)

and $2 : \frac{1}{3} = 6 : 1$ (multiplying both numbers by 3)

We can use this to simplify ratios containing fractions.

EXERCISE 12b

Express in their simplest form the ratios

a) $3 : \frac{1}{4}$ b) $\frac{2}{3} : \frac{4}{5}$

a) $3 : \frac{1}{4} = 12 : 1$ (multiplying both numbers by 4)

b) $\frac{2}{3} : \frac{4}{5} = \cancel{15}^{5} \times \frac{2}{\cancel{3}_1} : \cancel{15}^{3} \times \frac{4}{\cancel{5}_1}$ (multiplying both numbers by 15)

$= 10 : 12$

$= 5 : 6$

Use graph paper for questions 1 to 3.

Draw x and y axes, numbering each one from -20 to $+20$ at 5-unit intervals using 1 cm for 5 units on each axis.

1. a) Plot the points A(5, 5), B(18, 18), C(-10, -10).

   b) Draw the straight line through A, B and C.

   c) On your line mark the point D whose x coordinate is 10. What is the y coordinate of D?

   d) What is the connection between the x and y coordinates for each of the points A, B, C and D?

   e) Write down the equation of this line.

2. a) Plot the points A(1, 3), B(5, 15), C(-4, -12).

   b) Draw the straight line through A, B and C.

   c) On your line, mark the point D whose x coordinate is 2. What is the y coordinate of D?

   d) What is the connection between the x and y coordinates of each of the points A, B, C and D?

   e) Write down the equation of this line.

3. a) Plot the points A(10, -10), B(20, -20), C(-5, 5).

   b) Draw the straight line through A, B and C.

   c) On your line, mark the point D whose x coordinate is 5. What is the y coordinate of D?

   d) What is the connection between the x and y coordinates for each of the points A, B, C and D?

   e) Write down the equation of this line.

DRAWING A LINE FROM ITS EQUATION

Suppose that we want to draw the line whose equation is $y = 2x$.

For any point on this line,

   the y coordinate is twice the x coordinate.

Therefore if we choose a value of x, we can find the corresponding value of y.

We need only two points to draw a straight line, but it is sensible to use a third point as a check.

Express the following ratios in their simplest forms:

1. $5 : \frac{1}{3}$ <u>**6.**</u> $\frac{7}{12} : \frac{5}{6}$

2. $2 : \frac{1}{4}$ <u>**7.**</u> $\frac{5}{4} : \frac{6}{7}$

3. $\frac{1}{2} : \frac{1}{3}$ **8.** $3 : \frac{4}{3}$

4. $\frac{3}{4} : \frac{1}{4}$ **9.** $2\frac{2}{3} : 1\frac{1}{6}$

5. $\frac{1}{3} : \frac{3}{4}$ <u>**10.**</u> $\frac{2}{3} : \frac{7}{15}$

11. $24 : 15 : 9$ **16.** $\frac{1}{4} : \frac{1}{5}$

12. $\frac{4}{9} : \frac{2}{3}$ <u>**17.**</u> $1\frac{1}{2} : 3 : 4\frac{1}{2}$

13. $4 : \frac{9}{10}$ <u>**18.**</u> $6 : 4\frac{1}{2}$

14. $\frac{4}{5} : 6$ **19.** $\frac{1}{6} : \frac{1}{8} : \frac{1}{12}$

15. $7\frac{1}{2} : 9\frac{1}{2}$ <u>**20.**</u> $6 : 8 : 12$

RELATIVE SIZES

EXERCISE 12c

Which ratio is the larger, $6 : 5$ or $7 : 6$?

(We need to compare the sizes of $\frac{6}{5}$ and $\frac{7}{6}$ so we express both with the same denominator.)

$$\frac{6}{5} = \frac{36}{30} \quad \text{and} \quad \frac{7}{6} = \frac{35}{30}$$

so $6 : 5$ is larger than $7 : 6$

<u>**1.**</u> Which ratio is the larger, $5 : 7$ or $2 : 3$?

<u>**2.**</u> Which ratio is the smaller, $7 : 4$ or $13 : 8$?

<u>**3.**</u> Which ratio is the larger, $\frac{5}{8}$ or $\frac{7}{12}$?

<u>**4.**</u> Which ratio is the smaller, $\frac{3}{4}$ or $\frac{7}{10}$?

Which of the ratios $4 : 6$, $\frac{3}{4} : 1$, $12 : 16$ are equal to one another?

$4 : 6 = 2 : 3$ $\frac{3}{4} : 1 = 3 : 4$ $12 : 16 = 3 : 4$

so $\frac{3}{4} : 1 = 12 : 16$

In the following sets of ratios some are equal to one another. In each question identify the equal ratios.

5. $6 : 8$, $24 : 32$, $\frac{3}{4} : 1$

6. $10 : 24$, $\frac{5}{9} : \frac{4}{5}$, $\frac{5}{9} : \frac{4}{3}$

7. $8 : 64$, $2 : 14$, $\frac{1}{16} : \frac{1}{2}$

8. $\frac{2}{3} : 3$, $4 : 18$, $2 : 6$

PROBLEMS

EXERCISE 12d

A family has 12 pets of which 6 are cats or kittens, 2 are dogs and the rest are birds. Find the ratio of the numbers of a) birds to dogs   b) birds to pets.

There are 4 birds:

a)   Number of birds : number of dogs $= 4 : 2$

$= 2 : 1$

b)   Number of birds : number of pets $= 4 : 12$

$= 1 : 3$

In each question give your answer in its simplest form.

1. A couple have 6 grandsons and 4 granddaughters. Find
   a) the ratio of the number of grandsons to that of granddaughters
   b) the ratio of the number of granddaughters to that of grandchildren.

2. Square A has side 6 cm and square B has side 8 cm. Find the ratio of
   a) the length of the side of square A to the length of the side of square B
   b) the area of square A to the area of square B.

3. Tom walks 2 km to school in 40 minutes and John cycles 5 km to school in 15 minutes. Find the ratio of a) Tom's distance to John's distance   b) Tom's time to John's time.

4. Mary has 18 sweets and Jane has 12. As Mary has 6 sweets more than Jane she tries to even things out by giving Jane 6 sweets. What is the ratio of the number of sweets Mary has to the number Jane has a) to start with   b) to end with?

5. If $p : q = 2 : 3$, find the ratio $6p : 2q$

6. Rectangle A has length 12 cm and width 6 cm while rectangle B has length 8 cm and width 5 cm. Find the ratio of

a) the length of A to the length of B

b) the area of A to the area of B

c) the perimeter of A to the perimeter of B

d) the size of an angle of A to the size of an angle of B.

7. A triangle has sides of lengths 3.2 cm, 4.8 cm and 3.6 cm. Find the ratio of the lengths of the sides to one another.

8. Two angles of a triangle are 54° and 72°. Find the ratio of the size of the third angle to the sum of the first two.

9. For a school fete, Mrs Jones and Mrs Brown make marmalade in 1 lb jars. Mrs Jones makes 5 jars of lemon marmalade and 3 jars of orange. Mrs Brown makes 7 jars of lemon marmalade and 5 of grapefruit. Find the ratio of the numbers of jars of

a) lemon to orange to grapefruit

b) Mrs Jones' to Mrs Brown's marmalade

c) Mrs Jones' lemon to orange.

FINDING MISSING QUANTITIES

Some missing numbers are fairly obvious.

EXERCISE 12e

Find the missing numbers in the following ratios:

a) $6 : 5 = \quad : 10$ b) $\frac{4}{3} = \frac{}{9} = \frac{24}{}$

a) $6 : 5 = 12 : 10$

b) $\frac{4}{3} = \frac{12}{9} = \frac{24}{18}$

Find the missing numbers in the following ratios:

1. $2 : 5 = 4 :$

2. $\quad : 6 = 12 : 18$

3. $24 : 14 = 12 :$

4. $\frac{6}{} = \frac{9}{3}$

5. $3 : \quad = 12 : 32$

6. $\quad : 15 = 8 : 10$

7. $9 : 6 = \quad : 4$

8. $\frac{}{4} = \frac{15}{10}$

9. $\frac{6}{8} = \frac{}{12}$

10. $6 : 9 = 8 :$

Some missing numbers are not so obvious.

Find the missing numbers in

a) $: 4 = 3 : 5$ b) $6 : \ = 5 : 3$

(Fill the gap with an x to start with.)

a) $x : 4 = 3 : 5$ b) $6 : x = 5 : 3$

$\quad \dfrac{x}{4} = \dfrac{3}{5}$ or $x : 6 = 3 : 5$

$\quad 4 \times \dfrac{x}{4} = 4 \times \dfrac{3}{5}$ $\dfrac{x}{6} = \dfrac{3}{5}$

$\quad\quad x = \dfrac{12}{5}$ $6 \times \dfrac{x}{6} = 6 \times \dfrac{3}{5}$

$\quad\quad = 2\dfrac{2}{5}$ $x = \dfrac{18}{5}$

$2\dfrac{2}{5} : 4 = 3 : 5$ $= 3\dfrac{3}{5}$

$6 : 3\dfrac{3}{5} = 5 : 3$

11.   Use this method to repeat questions 6, 7 and 10.

Find x in questions 12 to 23.

12. $\dfrac{x}{3} = \dfrac{4}{5}$ **18.** $3 : 5 = x : 6$

13. $\dfrac{x}{4} = \dfrac{1}{3}$ **19.** $7 : 3 = 3 : x$

14. $x : 7 = 3 : 4$ **20.** $3 : x = 2 : 5$

15. $x : 5 = 4 : 3$ **21.** $5 : 1 = 3 : x$

16. $x : 4 = 1 : 3$ **22.** $6 : 5 = 12 : x$

17. $4 : x = 3 : 5$ **23.** $x : 3 = 7 : 15$

Find the missing numbers in questions 24 to 33.

24. $: 9 = 3 : 5$ **29.** $9 : 5 = \ : 4$

25. $: 3 = 5 : 2$ **30.** $10 : 3 = \ : 5$

26. $: 5 = 3 : 4$ **31.** $4 : 3 = 5 :$

27. $3 : \ = 5 : 1$ **32.** $: 6 = 5 : 8$

28. $4 : \ = 6 : 5$ **33.** $12 : \ = 10 : 3$

PROBLEMS

EXERCISE 12f

Two speeds are in the ratio $12 : 5$. If the first speed is $8\,\text{km/h}$, what is the second speed?

Let the second speed be $x\,\text{km/h}$. Then $8 : x = 12 : 5$

$$\frac{x}{8} = \frac{5}{12}$$

$$\cancel{8} \times \frac{x}{\cancel{8}} = \cancel{8} \times \frac{5}{\cancel{12}}$$

$$x = \frac{10}{3}$$

$$= 3\tfrac{1}{3}$$

The second speed is $3\tfrac{1}{3}\,\text{km/h}$

1. The ratio of the amount of money in David's pocket to that in Indira's pocket is $9 : 10$. Indira has $25\,\text{p}$. How much has David got?

2. Two lengths are in the ratio $3 : 7$. The second length is $42\,\text{cm}$. Find the first length.

3. If the ratio in question 2 were $7 : 3$, what would the first length be?

4. In a rectangle, the ratio of length to width is $9 : 4$. The length is $24\,\text{cm}$. Find the width.

5. The ratio of the perimeter of a triangle to its shortest side is $10 : 3$. The perimeter is $35\,\text{cm}$. What is the length of the shortest side?

6. A length, originally $6\,\text{cm}$, is increased so that the ratio of the new length to the old length is $9 : 2$. What is the new length?

7. A class is making a model of the school building and the ratio of the lengths of the model to the lengths of the actual building is $1 : 20$. The gym is $6\,\text{m}$ high. How high, in centimetres, should the model of the gym be?

8. The ratio of lengths of a model boat to those of the actual boat is $3 : 50$. Find the length of the actual boat if the model is $72\,\text{cm}$ long.

DIVISION IN A GIVEN RATIO

EXERCISE 12g

Share £60 between Anne and John so that Anne's share and John's share are in the ratio 3 : 2.

Anne has 3 portions and John has 2 portions so they have 5 portions between them.

$$\therefore \quad \text{Anne's share} = \frac{3}{5} \text{ of } £60$$
$$= £\frac{3}{\underset{1}{\cancel{5}}} \times \cancel{60}^{12}$$
$$= £36$$
$$\text{John's share} = £\frac{2}{\underset{1}{\cancel{5}}} \times \cancel{60}^{12}$$
$$= £24$$

Check: £36 + £24 = £60

1. Divide 80 p into two parts in the ratio 3 : 2.

2. Divide 32 cm into two parts in the ratio 3 : 5.

3. Divide £45 into two shares in the ratio 4 : 5.

4. Dick and Tom share the contents of a bag of peanuts between them in the ratio 3 : 5. If there are 40 peanuts, how many do they each get?

5. Mary is 10 years old and Eleanor is 15 years old. Divide 75 p between them in the ratio of their ages.

6. In a class of 30 pupils the ratio of the number of boys to the number of girls is 7 : 8. How many girls are there?

7. Divide £20 into two parts in the ratio 1 : 7.

8. In a garden the ratio of the area of lawn to the area of flowerbed is 12 : 5. If the total area is 357 m², find the area of
a) the lawn   b) the flowerbed.

9. In a bowl containing oranges and apples, the ratio of the numbers of oranges to apples is 4 : 3. If there are 28 fruit altogether, how many apples are there?

Divide 6 m into three parts in the ratio $3 : 7 : 2$.

There are 12 portions (that is, $3+7+2$)

$$\text{First part} = \frac{3}{\cancel{12}_1} \times \cancel{600}^{50} \text{cm}$$

$$= 150 \, \text{cm}$$

$$\text{Second part} = \frac{7}{\cancel{12}_1} \times \cancel{600}^{50} \text{cm}$$

$$= 350 \, \text{cm}$$

$$\text{Third part} = \frac{2}{\cancel{12}_1} \times \cancel{600}^{50} \text{cm}$$

$$= 100 \, \text{cm}$$

Check: $150+350+100 \, \text{cm} = 600 \, \text{cm} = 6 \, \text{m}$

10. Divide £26 amongst three people so that their shares are in the ratio $4 : 5 : 4$.

11. The perimeter of a triangle is 24 cm and the lengths of the sides are in the ratio $3 : 4 : 5$. Find the lengths of the three sides.

12. In a garden, the ratio of the areas of lawn to beds to paths is $3 : 1 : \frac{1}{2}$. Find the three areas if the total area is $63 \, \text{m}^2$.

MAP RATIO (OR REPRESENTATIVE FRACTION)

The Map Ratio of a map is the ratio of a length on the map to the length it represents on the ground. This ratio or fraction is given on most maps in addition to the scale. It is sometimes called the Representative Fraction of the map, or RF for short.

If two villages are 6 km apart and on the map this distance is represented by 6 cm, then the ratio is

$$6 \, \text{cm} : 6 \, \text{km} = 6 \, \text{cm} : 600\,000 \, \text{cm}$$

$$= 1 : 100\,000$$

so the map ratio is $1 : 100\,000$ or $\dfrac{1}{100\,000}$

Any length on the ground is 100 000 times the corresponding length on the map.

EXERCISE 12h

> Find the map ratio of a map if 12 km is represented by 1.2 cm on the map.
>
> $$RF = 1.2 \, cm : 12 \, km$$
> $$= 1.2 \, cm : 1\,200\,000 \, cm$$
> $$= 12 : 12\,000\,000 \quad \text{(multiplying both numbers by 10)}$$
> $$= 1 : 1\,000\,000 \quad \text{(dividing both numbers by 12)}$$

Find the map ratio of the maps in the following questions:

1. 2 cm on the map represents 1 km

2. The scale of the map is 1 cm to 5 km

3. 10 km is represented by 10 cm on the map

4. 3.2 cm on the map represents 16 km

5. $\frac{1}{2}$ cm on the map represents 500 m

6. 100 km is represented by 5 cm on the map

> If the map ratio is 1 : 5000 and the distance between two points on the map is 12 cm, find the actual distance between the two points.
>
> Let the actual distance be x cm.
>
> Then
> $$12 : x = 1 : 5000$$
> $$\text{or} \quad x : 12 = 5000 : 1$$
> $$\frac{x}{12} = \frac{5000}{1}$$
> $$12x \frac{x}{12} = 12 \times \frac{5000}{1}$$
> $$x = 60\,000$$
>
> The actual distance is 60 000 cm, that is, 600 m.
>
> *Alternative method:*
>   1 cm on the map represents 5000 cm on the ground.
>
> 12 cm on the map represents 12×5000 cm on the ground, i.e. 60 000 cm = 600 m

7. The map ratio of a map is 1 : 50 000. The distance between A and B on the map is 6 cm. What is the true distance between A and B?

8. The map ratio of a map is 1 : 1000. A length on the map is 7 cm. What real length does this represent?

9. The map ratio of a map is 1 : 10 000. Find the actual length represented by 2 cm.

10. The map ratio of a map is 1 : 200 000. The distance between two towns is 20 km. What is this in centimetres? Find the distance on the map between the points representing the towns.

11. The map ratio of a map is 1 : 2 000 000. Find the distance on the map which represents an actual distance of 36 km.

PROPORTION

When comparing quantities, words other than ratio are sometimes used. If two varying quantities are *directly proportional* they are always in the same ratio.

Sometimes it is obvious that two quantities are directly proportional, e.g. the cost of buying oranges is proportional to the number of oranges bought. In cases like this you would be expected to know that the quantities are in direct proportion.

EXERCISE 12i

A book of 250 pages is 1.5 cm thick (not counting the covers).

a) How thick is a book of 400 pages?

b) How many pages are there in a book 2.7 cm thick?

Method 1 (using algebra):

a) If the second book is x cm thick, then $\dfrac{x}{1.5} = \dfrac{400}{250}$

$$\overset{1}{\cancel{1.5}} \times \frac{x}{\cancel{1.5}} = \overset{0.3}{\cancel{1.5}} \times \frac{\overset{8}{\cancel{400}}}{\underset{5}{\cancel{250}}}$$

$$x = 2.4$$

The second book is 2.4 cm thick.

b) The third book has y pages, so $\dfrac{y}{250} = \dfrac{2.7}{1.5}$

$$\cancel{250} \times \frac{y}{\cancel{250}} = \overset{50}{\cancel{250}} \times \frac{\overset{9}{\cancel{27}}}{\underset{5}{\cancel{15}}}$$

$$y = 450$$

The third book has 450 pages.

Method 2 (*unitary method*):

a) 250 pages are 15 mm thick

1 page is $\frac{15}{250}$ mm thick

so 400 pages are $\frac{15}{250} \times 400$ mm thick

that is, 24 mm or 2.4 cm thick

b) 15 mm contains 250 pages

1 mm contains $\frac{250}{15}$ pages

so 27 mm contains $\frac{250}{15} \times 27$ pages

that is, 450 pages

1. Sam covers 9 m when he walks 12 paces. How far does he travel when he walks 16 paces?

2. I can buy 24 bottles of a cold drink for £8 when buying in bulk. How many bottles can I buy at the same rate for £12?

3. If 64 seedlings are allowed 24 cm^2 of space, how much space should be allowed for 48 seedlings? How many seedlings can be planted in 27 cm^2?

4. A ream (500 sheets) of paper is 6 cm thick. How thick a pile would 300 sheets make?

5. At a school picnic 15 sandwiches are provided for every 8 children. How many sandwiches are needed for 56 children?

Beware: some of the quantities in the following questions are not in direct proportion. Some questions need a different method and some cannot be answered at all from the given information.

A family with two pets spends £1.50 a week on pet food. If the family gets a third pet, how much a week will be spent on pet food?

We are not told what sort of animals the pets are. Different animals eat different types and quantities of food so the amount spent is not in proportion to the number of pets.

6. Two tea towels dry on a clothes line in 2 hours. How long would 5 tea towels take to dry?

7. Three bricklayers build a wall in 6 hours. How long would two bricklayers take to build the wall working at the same rate?

8. House contents insurance is charged at the rate of £3.50 per thousand pounds worth of the contents. How much is the insurance if the contents are worth £3400?

9. If the insurance paid on the contents of a house is £33.60, at the rate of £4 per thousand pounds worth, what are the house contents worth?

10. It takes Margaret 45 minutes to walk 4 km. How long would it take her to walk 5 km at the same speed? How far would she go in 1 hour?

11. It takes a gardener 45 minutes to dig a flower bed of area $7.5 \, m^2$. If he digs at the same rate, how long does he take to dig $9 \, m^2$?

12. Fencing costs £2.40 per 1.8 m length. How much would 7.5 m cost?

13. Mrs Brown and Mrs Jones make 4 dozen sandwiches in half an hour in Mrs Jones' small kitchen. If they had 30 friends in to help, how many sandwiches could be made in the same time?

14. A recipe for 12 scones requires 2 teaspoons of baking powder and 240 g of flour. If a larger number of scones are made, using 540 g of flour, how much baking powder is needed?

MIXED EXERCISES

EXERCISE 12j

1. Express the ratio 96 : 216 in its simplest form.

2. Simplify the ratio $\frac{1}{4} : \frac{2}{5}$

3. Divide £100 into three parts in the ratio 10 : 13 : 2

4. Two cubes have edges of lengths 8 cm and 12 cm. Find the ratio of a) the lengths of their edges   b) their volumes.

5. Find the missing number in the ratio   : 18 = 11 : 24

6. What does 1 cm represent on a map with map ratio 1 : 10 000?

7. If $x : y = 3 : 4$, find the ratio $4x : 3y$

8. It costs £4.50 to feed a dog for 12 days. At the same rate, how much will have to be spent to feed it for 35 days?

EXERCISE 12k **1.** Express the ratio $10\,\text{mm}^2 : 1\,\text{cm}^2$ in its simplest form.

2. Simplify the ratio $\frac{7}{8} : \frac{3}{4}$

3. Adrian has 24p and Brian has 36p. Give the ratio of the amount of Adrian's money to the total amount of money.

4. Which ratio is the larger, $16:13$ or $9:7$?

5. What is the map ratio of a map with a scale of 1 cm to 5 km?

6. Find the missing number in the ratio $7:12 = \quad :9$

7. Share £26 amongst three people in the ratio $6:3:4$

8. The ratio of boys to girls in a school is $10:9$. There are 459 girls. How many boys are there?

EXERCISE 12l **1.** Express the ratio $1028:576$ in its simplest form.

2. Which ratio is the smaller, $32:24$ or $30:22$?

3. An alloy is made of copper and zinc in the ratio $11:2$. How much zinc does 65 kg of alloy contain?

4. Increase a length of 24 m so that the ratio of the new length to the old length is $11:8$

5. Anne has twice as many crayons as Martin, who has three times as many as Susan. Give the ratio of the number of crayons owned by the three children.

6. The map ratio of a map is $1:50\,000$. Find the length on the ground represented by 6.4 cm on the map.

7. Simplify the ratio $\frac{13}{12} : \frac{5}{21}$

8. Carpet to cover a floor of area $15\,\text{m}^2$ costs £110. How much would you expect to pay for a similar carpet measuring 5 m by 4.2 m?

13 ENLARGEMENTS

ENLARGEMENTS

All the transformations we have used so far (i.e. reflections, translations and rotations) have moved the object and perhaps turned it over to produce the image, but its shape and size have not changed. Next we come to a transformation which keeps the shape but alters the size.

Think of the picture thrown on the screen when a slide projector is used.

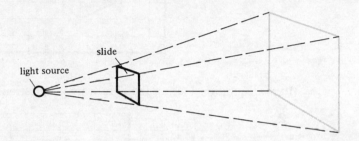

The picture on the screen is the same as that on the slide but it is very much bigger.

We can use the same idea to enlarge any shape.

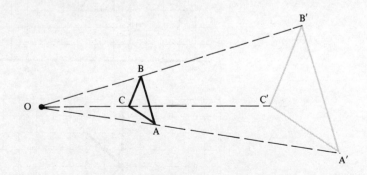

△A′B′C′ is the image of △ABC under an *enlargement*, *centre O*.

O is the *centre of enlargement*.

We call the dotted lines *guidelines*.

199

CENTRE OF ENLARGEMENT

In all these questions, one triangle is an enlargement of the other.

EXERCISE 13a   **1.** Copy the diagram using 1 cm to 1 unit. Draw P'P, Q'Q and R'R and continue all three lines until they meet.
The point where the lines meet is called the centre of enlargement.
Give the coordinates of the centre of enlargement.

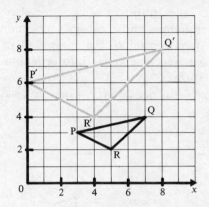

Repeat question 1 using the diagrams in questions 2 and 3.

2.

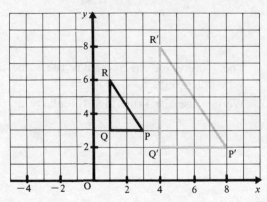

3.

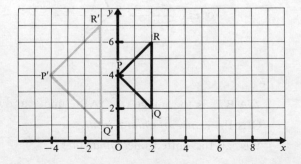

4. In questions 1 to 3, name pairs of lines that are parallel.

5. Draw axes for x and y from 0 to 9 using 1 cm as 1 unit.
Draw $\triangle$ABC: A(2, 3), B(4, 1), C(5, 4).
Draw $\triangle$A′B′C′: A′(2, 5), B′(6, 1), C′(8, 7).
Draw A′A, B′B and C′C and extend these lines until they meet.

a) Give the coordinates of the centre of enlargement.

b) Measure the sides and angles of the two triangles. What do you notice?

6. Repeat question 5 with $\triangle$ABC: A(8, 4), B(6, 6), C 6, 4) and $\triangle$A′B′C′: A′(6, 2), B′(0, 8), C′(0, 2)

7. Draw axes for x and y from 0 to 10 using 1 cm as 1 unit.
Draw $\triangle$XYZ with X(8, 2), Y(6, 6) and Z(5, 3)
and $\triangle$X′Y′Z′ with X′(6, 2), Y′(2, 10) and Z′(0, 4).
Find the centre of enlargement and label it P.
Measure PX, PX′, PY, PY′, PZ, PZ′. What do you notice?

The centre of enlargement can be anywhere, including a point inside the object or a point on the object.

The centres of enlargement in the diagrams below are marked with a cross.

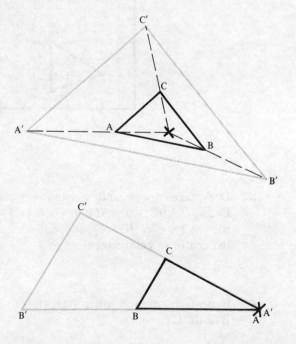

EXERCISE 13b   **1.**   Copy the diagram using 1 cm as 1 unit. Draw A'A, B'B and C'C and extend the lines until they meet. Give the coordinates of the centre of enlargement.

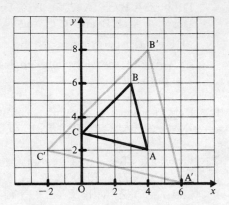

2.   In the diagram below which point is the centre of enlargement?

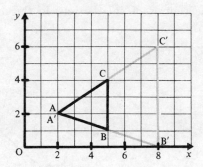

3.   Draw axes for x and y from -3 to 10 using 1 cm as 1 unit. Draw $\triangle ABC$ with A(4, 0), B(4, 4) and C(0, 2). Draw $\triangle A'B'C'$ with A'(5, -2), B'(5, 6) and C'(-3, 2). Find the coordinates of the centre of enlargement.

4.   Repeat question 3 with A(1, 4), B(5, 2), C(5, 5) and A'(-3, 6), B'(9, 0), C'(9, 9).

SCALE FACTORS

If we measure the lengths of the sides of the two triangles PQR and P'Q'R' and compare them, we find that the lengths of the sides of △P'Q'R' are three times those of △PQR.

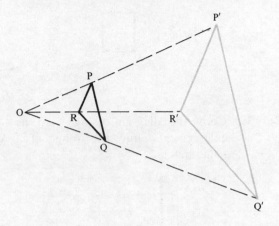

We say that △P'Q'R' is the image of △PQR under an enlargement, centre O, with *scale factor 3*.

FINDING AN IMAGE UNDER ENLARGEMENT

If we measure OR and OR' in the diagram above, we find R' is three times as far from O as R is. This enables us to work out a method for enlarging an object with a given centre of enlargement (say O) and a given scale factor (say 3).

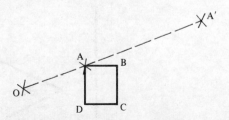

Measure OA. Multiply it by 3. Mark A' on the guideline three times as far from O as A is.

$$OA' = 3 \times OA$$

Repeat for B and the other vertices of ABCD.

Then A′B′C′D′ is the image of ABCD. To check, measure A′B′ and AB. A′B′ should be three times as large as AB.

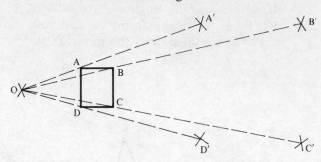

EXERCISE 13c **1.** Copy the diagram using 1 cm as 1 unit. P is the centre of enlargement. Draw the image of △ABC under an enlargement scale factor 2.

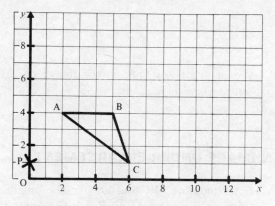

2. Repeat question 1 using this diagram.

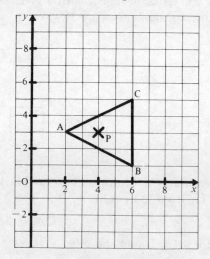

In questions 3 to 6, draw axes for x and y from 0 to 10, using 1 cm as 1 unit. In each case, find the image A'B'C' of △ABC using the given enlargement. Check by measuring the lengths of the sides of the two triangles.

3.   △ABC: A(3, 3), B(6, 2), C(5, 6).
Enlargement with centre (5, 4), and scale factor 2.

4.   △ABC: A(1, 2), B(3, 2), C(1, 5).
Enlargement with centre (0, 0) and scale factor 2.
What do you notice about the coordinates of A' compared with those of A?

5.   △ABC: A(2, 1), B(4, 1) C(3, 4).
Enlargement with centre (1, 1) and scale factor 3.

6.   △ABC: A(1, 2), B(7, 2), C(1, 6).
Enlargement with centre (1, 2) and scale factor $1\frac{1}{2}$.

7.   On plain paper, mark a point P near the left-hand edge. Draw a small object (a pin man perhaps, or a square house) between P and the middle of the page. Using the method of enlargement, draw the image of the object with centre P and scale factor 2.

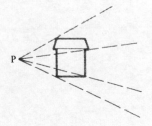

8.   Repeat question 7 with other objects and other scale factors. Think carefully about the space you will need for the image.

9.   Draw axes for x and y from 0 to 10 using 1 cm as 1 unit. Draw △ABC with A(2, 2), B(5, 1) and C(3, 4). Taking the origin as the centre of enlargement and a scale factor of 2, draw the image of △ABC by counting squares and without drawing the guidelines.

10.   Draw axes for x and y from 0 to 8 using 1 cm as 1 unit. Draw △ABC with A(1, 2), B(5, 2) and C(2, 5). Taking (3, 2) as the centre of enlargement and a scale factor of 2, draw the image of △ABC by counting squares and without drawing the guidelines.

FRACTIONAL SCALE FACTORS

We can reverse the process of enlargement and shrink or reduce the object, producing a smaller image. If the lengths of the image are one-third of the lengths of the object then the scale factor is $\frac{1}{3}$.

There is no satisfactory word to cover both enlargement and shrinking (some people use "dilation" and some "scaling") so *enlargement* tends to be used for both. You can tell one from the other by looking at the size of the scale factor. A scale factor smaller than 1 gives a smaller image while a scale factor greater than 1 gives a larger image.

EXERCISE 13d In questions 1 to 4, $\triangle A'B'C'$ is the image of $\triangle ABC$. Give the centre of enlargement and the scale factor.

1.

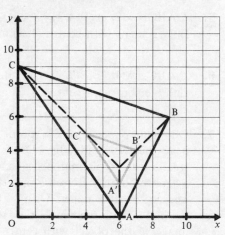

2.

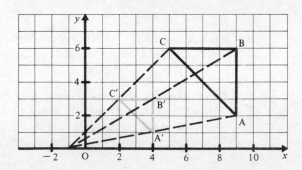

3. Draw axes for x and y from -2 to 8 using 1 cm as 1 unit. Draw $\triangle ABC$ with $A(-1, 4)$, $B(5, 1)$ and $C(5, 7)$, and $\triangle A'B'C'$ with $A'(2, 4)$, $B'(4, 3)$ and $C'(4, 5)$.

4. Draw axes for x and y from 0 to 9 using 1 cm as 1 unit. Draw $\triangle ABC$ with A(1, 2), B(9, 2) and C(9, 6), and $\triangle A'B'C'$ with A'(1, 2), B'(5, 2) and C'(5, 4).

In questions 5 and 6, draw axes for x and y from -1 to 11 using 1 cm as 1 unit. Find the image of $\triangle ABC$ under the given enlargement.

5. $\triangle ABC$: A(9, 1), B(11, 5), C(7, 7). Centre $(-1, 1)$, scale factor $\frac{1}{2}$.

6. $\triangle ABC$: A(4, 0), B(10, 9), C(1, 6). Centre $(4, 3)$, scale factor $\frac{1}{3}$.

NEGATIVE SCALE FACTORS

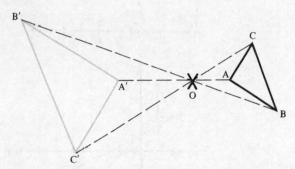

As you can see in the diagram above it is possible to produce an image twice the size of the object by drawing the guidelines backwards rather than forwards from the centre O. To show that we are going the opposite way we say that the scale factor is -2.

The image is the same shape but has been rotated through a half turn compared with the image produced by a scale factor of $+2$.

The following diagrams show enlargements which have scale factors of -3 and $+3$.

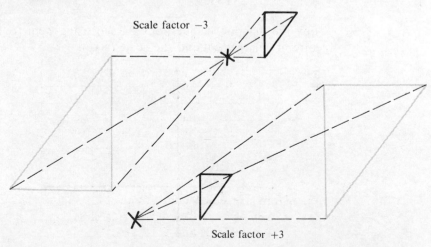

Scale factor -3

Scale factor $+3$

EXERCISE 13e In questions 1 and 2 give the centre of enlargement and the scale factor.

1.

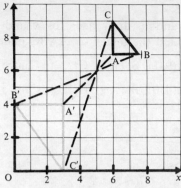

2.

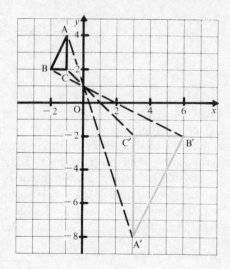

Copy the diagram in questions 3 to 6 using 1 cm to 1 unit. Find the centre of enlargement and the scale factor.

3.

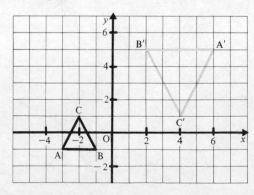

4.

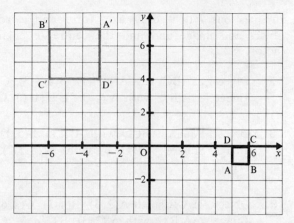

5.

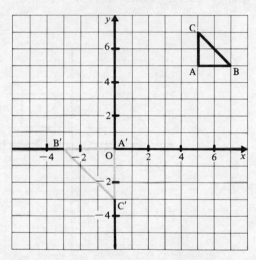

6.

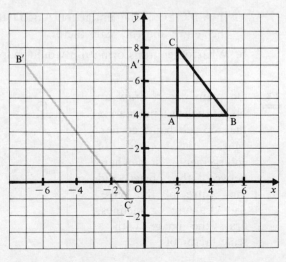

In questions 7 to 9, draw axes for x and y from -6 to 6. Draw the object and image and find the centre of enlargement and the scale factor.

7. Object: $\triangle ABC$ with $A(6, -1)$, $B(4, -3)$, $C(4, -1)$
Image: $\triangle A'B'C'$ with $A'(-3, 2)$, $B'(1, 6)$, $C'(1, 2)$

8. Object: Square ABCD with $A(1, 1)$, $B(5, 1)$, $C(5, -3)$, $D(1, -3)$
Image: Square $A'B'C'D'$ with $A'(-2, 2)$, $B'(-4, 2)$, $C'(-4, 4)$, $D'(-2, 4)$

9. Object: $\triangle ABC$ with $A(2, 3)$, $B(4, 3)$, $C(2, 6)$
Image: $\triangle A'B'C'$ with $A'(2, 3)$, $B'(-4, 3)$, $C'(2, -6)$

10.

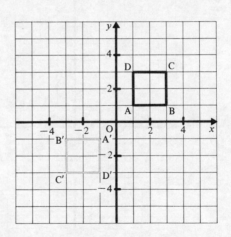

a) If $A'B'C'D'$ is the image of ABCD under enlargement, give the centre and the scale factor.

b) What other transformation would map ABCD to $A'B'C'D'$?

11. On plain paper, draw an object such as a pin man in the top left-hand corner. Mark the centre of enlargement somewhere between the object and the centre of the page. By drawing guidelines, draw the image with a scale factor of -2.

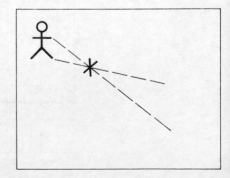

In questions 12 and 13, copy the diagrams and find the images of the triangles using P as the centre of enlargement and a scale factor of −2.

12.

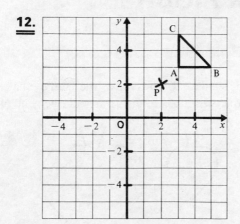

13.

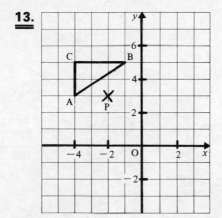

14. Draw axes for *x* from −10 to 4 and for *y* from −2 to 2.
Draw △ABC with A(2, 1), B(4, 1) and C(2, 2).
If the centre of enlargement is (1, 1) and the scale factor is −3, find the image of △ABC.

14 SIMILAR FIGURES

SIMILAR FIGURES

Two figures are similar if they are the same shape though not necessarily the same size. One figure is an enlargement of the other.

One may be turned round compared with the other.

One figure may be turned over compared with the other.

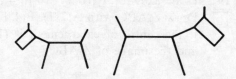

The following figures are not similar although their angles are equal.

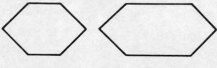

EXERCISE 14a State whether or not the pairs of figures in questions 1 to 10 are similar.

1.

6.

2.

7.

3.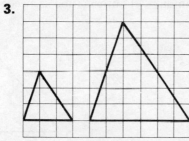

8.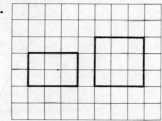

4.

9.

5.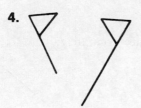

10.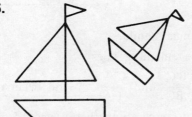

11. Which two rectangles are similar?

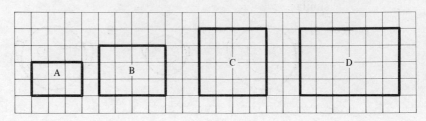

12. Draw your own pairs of figures and state whether or not they are similar. (The second figure may be turned round or over or both, compared with the first.)

SIMILAR TRIANGLES

Some of the easiest similar figures to deal with are triangles. This is because only a small amount of information is needed to prove them to be similar.

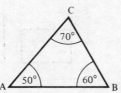

In these triangles the corresponding angles are equal and so the triangles are the same shape. One triangle is an enlargement of the other. These triangles are *similar*.

EXERCISE 14b   **1.** Draw the following triangles accurately:

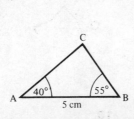
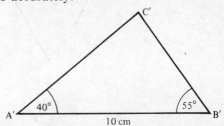

a) Are the triangles similar?

b) Measure the remaining sides.

c) Find $\dfrac{A'B'}{AB}$, $\dfrac{B'C'}{BC}$ and $\dfrac{C'A'}{CA}$ (as decimals if necessary)

d) What do you notice about the answers to part c)?

Repeat question 1 for the pairs of triangles in questions 2 to 5.

2.

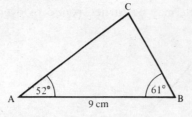

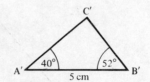

3.

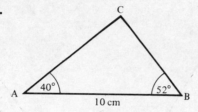

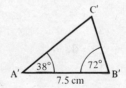

4.

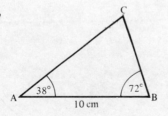

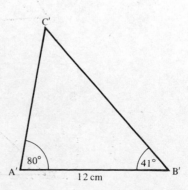

5.

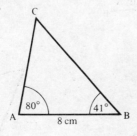

Sketch the following pairs of triangles and find the sizes of the missing angles. In each question state whether the two triangles are similar. (One triangle may be turned round or over compared with the other.)

6.

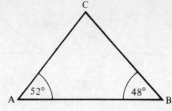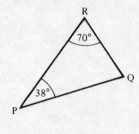

7.

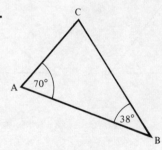

8.

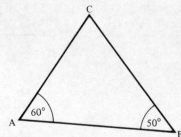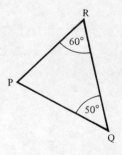

9.

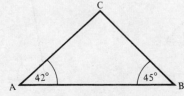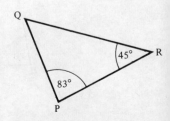

CORRESPONDING VERTICES

These two triangles are similar and we can see that X corresponds to A, Y to B and Z to C.

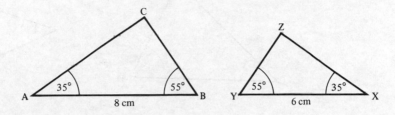

We can write: $\triangle$s $\dfrac{ABC}{XYZ}$ are similar

Make sure that X is written below A, Y below B and Z below C.

The pairs of corresponding sides are in the same ratio,

that is $\dfrac{AB}{XY} = \dfrac{BC}{YZ} = \dfrac{CA}{ZX}$

EXERCISE 14c

State whether triangles ABC and PQR are similar and if they are, give the ratios of the sides.

$\hat{Q} = 58°$ (angles of a triangle)

and $\hat{C} = 90°$

so $\triangle$s $\dfrac{ABC}{RQP}$ are similar

and $\dfrac{AB}{RQ} = \dfrac{BC}{QP} = \dfrac{CA}{PR}$

In questions 1 to 8, state whether the two triangles are similar and, if they are, give the ratios of the sides.

1.

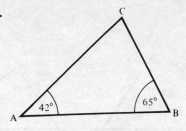
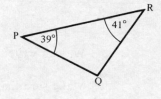

2.

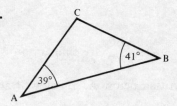
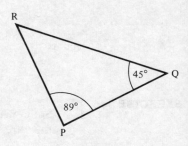

3.

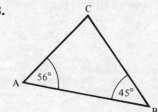
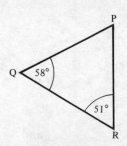

4.

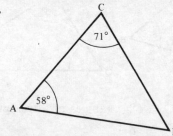

5. Use the triangles given in question 6 of Exercise 14b.

6. Use the triangles given in question 7 of Exercise 14b.

7. Use the triangles given in question 8 of Exercise 14b.

8. Use the triangles given in question 9 of Exercise 14b.

FINDING A MISSING LENGTH

EXERCISE 14d

State whether the two triangles are similar. If they are, find AB.

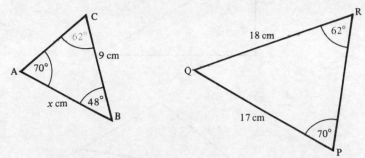

$\widehat{C} = 62°$ and $\widehat{Q} = 48°$ (angles of a triangle)

so $\triangle$s $\begin{matrix} ABC \\ PQR \end{matrix}$ are similar and $\dfrac{AB}{PQ} = \dfrac{BC}{QR} = \dfrac{CA}{RP}$

$$\frac{x}{17} = \frac{9}{18}$$

$$\cancel{17} \times \frac{x}{\cancel{17}} = \frac{\cancel{9}}{\cancel{18}_2} \times 17$$

$$x = \frac{17}{2} = 8.5$$

$$AB = 8.5\,cm$$

In questions 1 to 4, state whether pairs of triangles are similar. If they are, find the required side.

1. Find PR.

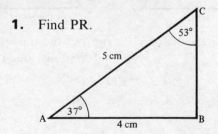

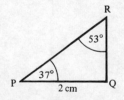

2. Find QR.

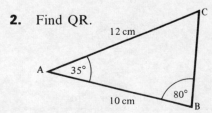

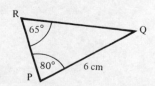

3. Find BC.

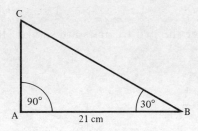

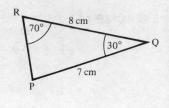

4. Find PR.

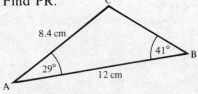

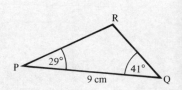

In some cases we do not need to know the sizes of the angles as long as we know that pairs of angles are equal. (Two pairs only are needed as the third pair must then be equal.)

In △s ABC and DEF, $\widehat{A} = \widehat{E}$ and $\widehat{B} = \widehat{D}$. AB = 4 cm, DE = 3 cm and AC = 6 cm. Find EF.

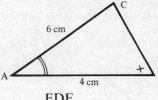

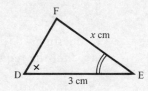

△s $\begin{matrix} EDF \\ ABC \end{matrix}$ are similar   (we put the triangle with the unknown side on top)

$$\frac{FE}{CA} = \frac{ED}{AB} = \frac{DF}{BC}$$

$$\frac{x}{6} = \frac{3}{4}$$

$$\overset{1}{\cancel{6}} \times \frac{x}{\cancel{6}_{1}} = \frac{3}{\cancel{4}_{2}} \times \cancel{6}^{3}$$

$$x = \frac{9}{2}$$

$$= 4.5$$

so   EF = 4.5 cm

5. In △s ABC and XYZ, $\widehat{A} = \widehat{X}$ and $\widehat{B} = \widehat{Y}$.
AB = 6cm, BC = 5cm and XY = 9cm. Find YZ.

6. In △s ABC and PQR, $\widehat{A} = \widehat{P}$ and $\widehat{C} = \widehat{R}$.
AB = 10cm, PQ = 12cm and QR = 9cm. Find BC.

7. In △s ABC and DEF, $\widehat{A} = \widehat{E}$ and $\widehat{B} = \widehat{F}$.
AB = 3cm, EF = 5cm and AC = 5cm. Find DE.

8. In △s ABC and PQR, $\widehat{A} = \widehat{Q}$ and $\widehat{C} = \widehat{R}$.
AC = 8cm, BC = 4cm and QR = 9cm. Find PR.

---

a) Show that triangles ABC and CDE are similar.

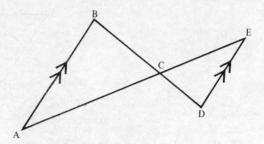

b) Given that AC = 15cm, CE = 9cm and DE = 8cm,
find AB.

a) $\widehat{A} = \widehat{E}$ (alternate angles, AB‖DE)

   $\widehat{B} = \widehat{D}$ (alternate angles, AB‖DE)

(Or we could use $B\widehat{C}A = E\widehat{C}D$ as these are vertically opposite angles.)

so △s $\dfrac{ABC}{EDC}$ are similar.

b)

$$\frac{AB}{ED} = \frac{BC}{DC} = \frac{CA}{CE}$$

$$\frac{x}{8} = \frac{15}{9}$$

$$\cancel{8} \times \frac{x}{\cancel{8}} = \frac{\cancelto{5}{15}}{\cancelto{3}{9}} \times 8$$

$$x = \frac{40}{3}$$

$$= 13\tfrac{1}{3}$$

AB = $13\tfrac{1}{3}$cm,   or 13.3cm   correct to 3 s.f.

9. a) Show that △s ABC and BDE are similar.

b) If AB = 6 cm, BD = 3 cm and DE = 2 cm, find BC.

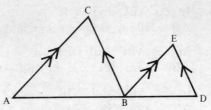

10. a) Show that △s ABC and CDE are similar.

b) If AB = 7 cm, BC = 6 cm, AC = 4 cm and CE = 6 cm, find CD and DE.

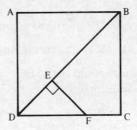

11. a) ABCD is a square. EF is at right angles to BD. Show that △s ABD and DEF are similar.

b) If AB = 10 cm, DB = 14.2 cm and DF = 7.1 cm, find EF.

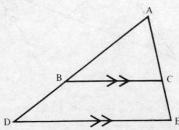

12. a) Show that △s ABC and ADE are similar. (Notice that $\hat{A}$ is *common* to both triangles.)

b) If AB = 10 cm, AD = 15 cm, BC = 12 cm and AC = 9 cm, find DE, AE and CE.

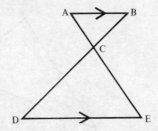

USING THE SCALE FACTOR TO FIND THE MISSING LENGTH

Sometimes the scale factor for enlarging one triangle into the other is very obvious and we can make use of this to save ourselves some work.

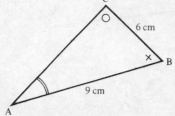 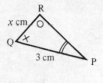

The two triangles above are similar and we can see that the scale factor for "enlarging" the first triangle into the second is $\frac{1}{3}$. We can say straightaway that x is $\frac{1}{3}$ of 6.

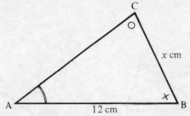 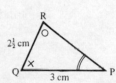

If we wish to find a length in the first triangle, we use the scale factor for enlarging the second triangle into the first.

The scale factor is 4 so $x = 4 \times 2\frac{1}{2} = 10$

EXERCISE 14e

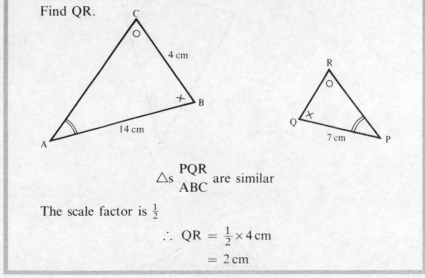

Find QR.

$\triangle$s $\begin{array}{c} \text{PQR} \\ \text{ABC} \end{array}$ are similar

The scale factor is $\frac{1}{2}$

$\therefore$ QR $= \frac{1}{2} \times 4\,$cm

$= 2\,$cm

1. Find BC.

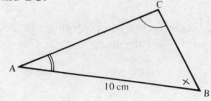

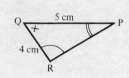

2. Find PR.

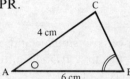

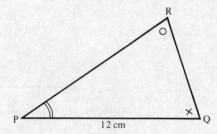

3. Find PR.

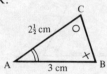

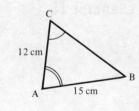

4. Find XY.

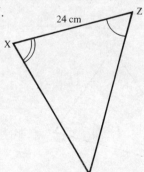

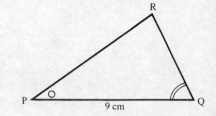

5. Find LN.

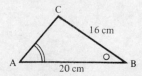

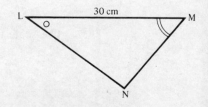

6. Find PQ.

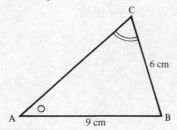 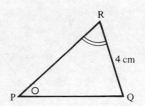

CORRESPONDING SIDES

If the three pairs of sides of two triangles are in the same ratio, then the triangles are similar and their corresponding angles are equal.

When finding the ratio of three sides give the ratio as a whole number or as a fraction in its lowest terms.

EXERCISE 14f

State whether triangles ABC and PQR are similar. Say which angle, if any, is equal to $\hat{A}$.

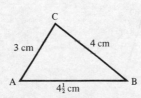 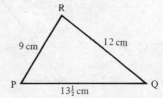

(Start with the shortest side of each triangle.)

$$\frac{PR}{AC} = \frac{9}{3} = 3$$

$$\frac{QR}{BC} = \frac{12}{4} = 3$$

$$\frac{PQ}{AB} = \frac{13\frac{1}{2}}{4\frac{1}{2}} = \frac{27}{9} = 3$$

i.e.
$$\frac{PR}{AC} = \frac{PQ}{AB} = \frac{QR}{BC}$$

so $\triangle$s $\dfrac{PQR}{ABC}$ are similar

$$\therefore \quad \hat{P} = \hat{A}$$

State whether the following pairs of triangles are similar. In each case say which angle, if any, is equal to $\hat{A}$.

1.

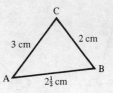 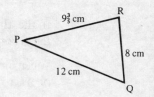

2.

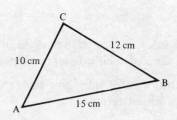 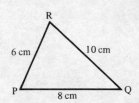

3.

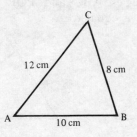 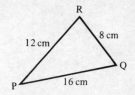

4.

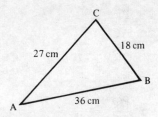

5.

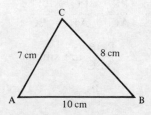 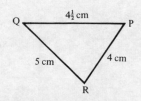

6.

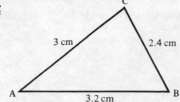

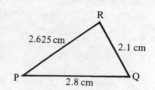

7. Are the triangles ABC and ADE similar?
Which angles are equal?
What can you say about lines BC and DE?

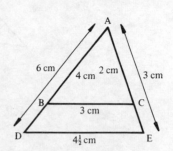

ONE PAIR OF EQUAL ANGLES AND TWO PAIRS OF SIDES

The third possible set of information about similar triangles concerns a pair of angles and the sides containing them.

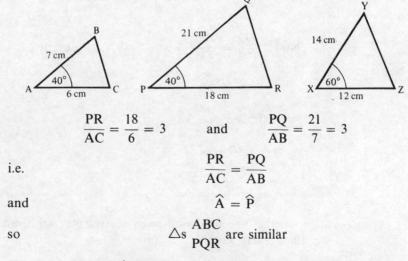

$$\frac{PR}{AC} = \frac{18}{6} = 3 \qquad \text{and} \qquad \frac{PQ}{AB} = \frac{21}{7} = 3$$

i.e.
$$\frac{PR}{AC} = \frac{PQ}{AB}$$

and
$$\hat{A} = \hat{P}$$

so
$$\triangle s \; \frac{ABC}{PQR} \; \text{are similar}$$

We can see that $\triangle PQR$ is an enlargement of $\triangle ABC$ and that the scale factor is 3. $\left(\text{It is given by } \dfrac{PQ}{AB}.\right)$

On the other hand $\triangle XYZ$ is a different shape from the other two and is not similar to either of them even though two pairs of sides are in the same ratio.

EXERCISE 14g

State whether triangles ABC and PQR are similar. If they are, find PQ.

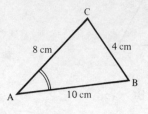

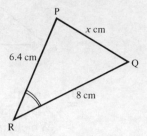

$$\frac{RP}{AC} = \frac{6.4}{8} = 0.8 \quad \text{(comparing the two shorter sides)}$$

$$\frac{RQ}{AB} = \frac{8}{10} = 0.8$$

$$\therefore \qquad \frac{RP}{AC} = \frac{RQ}{AB} \qquad \text{and} \qquad \widehat{A} = \widehat{R}$$

so $\qquad\qquad \triangle s \begin{array}{c} RQP \\ ABC \end{array}$ are similar

Now, $\dfrac{PQ}{CB} = \dfrac{RQ}{AB}$

$$\frac{x}{4} = \frac{8}{10}$$

$$^1\cancel{4} \times \frac{x}{\cancel{4}_1} = \frac{\cancel{8}^4}{\cancel{10}_5} \times 4$$

$$x = 3.2$$

$$PQ = 3.2\,\text{cm}$$

or: BC is half AC

so PQ is half PR

$\therefore$ PQ = 3.2 cm

State whether the following pairs of triangles are similar. If they are, find the missing lengths.

1.

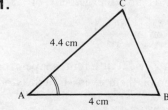

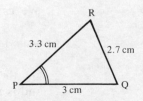

2.

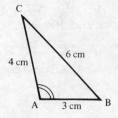

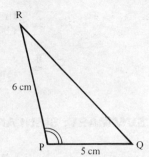

3.

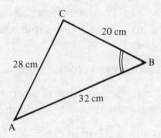

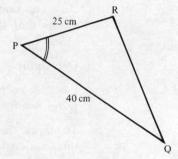

4.

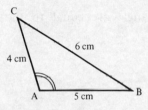

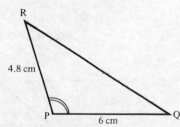

5.

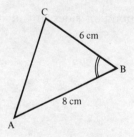

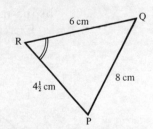

6.

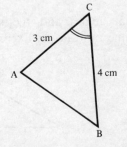

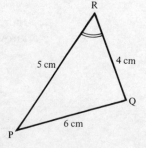

7. In △s ABC and PQR, $\widehat{A} = \widehat{P}$, AB = 8 cm, BC = 8.5 cm, CA = 6.5 cm, PQ = 4.8 cm and PR = 3.9 cm. Find QR.

8. In △s PQR and XYZ, $\widehat{P} = \widehat{X}$, PQ = 4 cm, PR = 3 cm, QR = $2\frac{1}{4}$ cm, XY = $5\frac{1}{3}$ cm and XZ = 4 cm. Find ZY.

SUMMARY: SIMILAR TRIANGLES

If two triangles are the same shape (but not necessarily the same size) they are said to be *similar*. This word, when used in mathematics, means that the triangles are *exactly* the same shape and not vaguely alike, as two sisters may be.

One triangle may be turned over or round compared with the other.

Pairs of corresponding sides are in the same ratio. This ratio is the *scale factor* for the enlargement of one triangle into the other.

To check that two triangles are similar we need to' show *one* of the three following sets of facts:

a) the angles of one triangle are equal to the angles of the other (as in Exercise 14c)

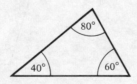

b) the three pairs of corresponding sides are in the same ratio (as in Exercise 14f)

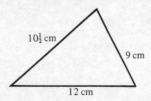

c) there is one pair of equal angles and the sides containing the known angles are in the same ratio (as in Exercise 14g).

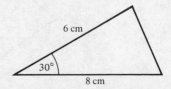

MIXED EXERCISE

EXERCISE 14h State whether or not the pairs of triangles in questions 1 to 10 are similar, giving your reasons. If they are similar, find the required side or angle.

1. Find BC.

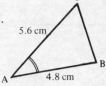
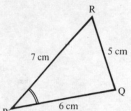

2. Find QR.

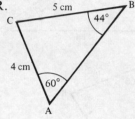
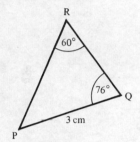

3. Find AC.

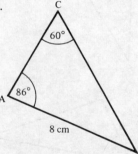
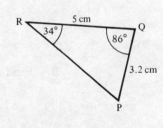

4. Find $\hat{Q}$.

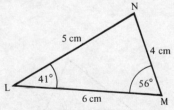
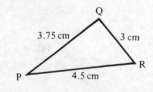

5. Find FE.

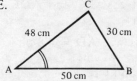
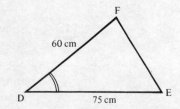

6. Find $\hat{X}$.

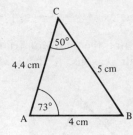
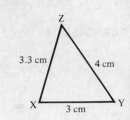

7. Find $\hat{P}$.

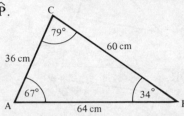
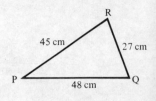

8. Find $\hat{Q}$.

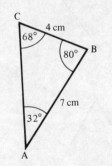
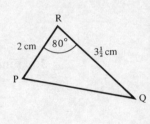

9. Find YZ.

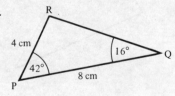
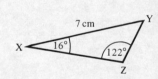

10. Find AC.

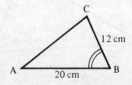
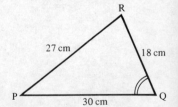

11. a) Show that △s ABC and ADE are similar.

b) AB = 3.6 cm,   AD = 4.8 cm and   AE = 4.2 cm.
Find AC and CE.

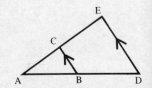

12.   a) Show that △s ABC and DEF are similar.

b) AB = 40 cm, BC = 52 cm and DE = 110 cm. Find EF.

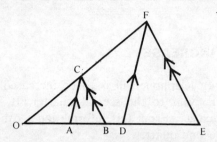

13.   In the figure below there are three overlapping triangles.

a) Show that △s ABC and ABD are similar.

b) Show that △s ABC and BDC are similar.

c) Are △s ABD and BDC similar?

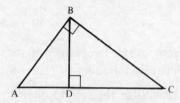

14.   A pole, AB, 2 m high, casts a shadow, AC, that is 3 m long. Another pole, PQ, casts a shadow 15 m long. How high is the second pole?

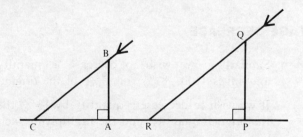

15.   The shadow of a 1 m stick held upright on the ground is 2.4 m long. How long a shadow would be cast by an 8 m telegraph pole?

16.   A slide measures 1.8 cm by 2.4 cm. A picture 90 cm by 120 cm is cast on the screen. On the slide, a house is 1.2 cm high. How high is the house in the picture on the screen?

15 PERCENTAGE INCREASE AND DECREASE

PERCENTAGE INCREASE

My telephone bill is to be increased by 8% from the first quarter of the year to the second quarter. It amounted to £64.50 for the first quarter. From this information I can find the value of the bill for the second quarter.

If £64.50 is increased by 8%, the increase is 8% of £64.50,

i.e.

$$£\frac{8}{100} \times 64.50 = £5.16$$

The bill for the second quarter is therefore

$$£64.50 + £5.16 = £69.66$$

The same result is obtained if we take the original sum to be 100%. The increased amount is $(100+8)\%$, or $\frac{108}{100}$, of the original sum,

i.e. the bill for the second quarter is $\qquad £\frac{108}{100} \times 64.50 = £69.66$

The quantity $\frac{108}{100}$ is called the multiplying factor and to increase a quantity by 12%, the multiplying factor would be $\frac{112}{100}$.

PERCENTAGE DECREASE

Similarly if we wish to decrease a quantity by 8%, the decreased amount is $(100-8)\%$, or $\frac{92}{100}$, of the original sum.

If we wish to decrease a quantity by 15%, the new quantity is 85% of the original quantity, and the multiplying factor is $\frac{85}{100}$.

EXERCISE 15a

If a number is increased by 40%, what percentage is the new number of the original number?

The new number is 140% of the original.

If a number is increased by the given percentage, what percentage is the new number of the original number?

1. 50%	**4.** 60%	**7.** 48%	**10.** $12\frac{1}{2}$%
2. 25%	**5.** 75%	**8.** 300%	**11.** 57%
3. 20%	**6.** 35%	**9.** 175%	**12.** 15%

What multiplying factor increases a number by 44%?

The multiplying factor is $\dfrac{100+44}{100} = \dfrac{144}{100}$

Give the multiplying factor which increases a number by:

13. 30%   **14.** 80%   **15.** 65%   **16.** 130%

If a number is decreased by 65%, what percentage is the new number of the original number?

The new number is 35% of the original.

If a number is decreased by the given percentage what percentage is the new number of the original number?

17. 50%	**20.** 85%	**23.** 4%	**26.** $33\frac{1}{3}$%
18. 25%	**21.** 35%	**24.** 66%	**27.** 53%
19. 70%	**22.** 42%	**25.** $62\frac{1}{2}$%	**28.** 10%

What multiplying factor decreases a number by 30%?

The multiplying factor is $\dfrac{100-30}{100} = \dfrac{70}{100}$

What multiplying factor decreases a number by:

29. 40%   **30.** 75%   **31.** 34%   **32.** 12%

> Increase 180 by 30%.
>
> The new value is 130% of the old
>
> i.e.   the new value is $\dfrac{130}{100} \times 180 = 234$

Increase:

33. 100 by 40%

34. 200 by 85%

35. 340 by 45%

36. 550 by 36%

37. 1600 by 73%

38. 745 by 14%

39. 64 by $62\frac{1}{2}$%

40. 111 by $66\frac{2}{3}$%

41. 145 by 120%

42. 644 by 275%

> Decrease 250 by 70%.
>
> The new value is 30% of the original value
>
> i.e.   the new value is $\dfrac{30}{100} \times 250 = 75$

Decrease:

43. 100 by 30%

44. 200 by 15%

45. 350 by 46%

46. 750 by 13%

47. 3400 by 28%

48. 3450 by 4%

49. 93 by $33\frac{1}{3}$%

50. 273 by $66\frac{2}{3}$%

51. 208 by $87\frac{1}{2}$%

52. 248 by $37\frac{1}{2}$%

PROBLEMS

EXERCISE 15b

1. A boy's weight increased by 15% between his fifteenth and sixteenth birthdays. If he weighed 55 kg on his fifteenth birthday, what did he weigh on his sixteenth birthday?

2. The water rates due on my house this year are 8% more than they were last year. Last year I paid £210. What must I pay this year?

3. There are 80 teachers in a school. It is anticipated that the number of staff next year will increase by 5%. How many staff should there be next year?

4. Pierre is 20% taller now than he was 2 years ago. If he was 150 cm tall then, how tall is he now?

5. A factory employs 220 workers. Next year this number will increase by 15%. How many extra workers will be taken on?

6. A bathroom suite is priced at £650 plus value added tax (VAT) at 15%. How much does the suite actually cost the customer?

7. An LP record costs £7 plus value added tax at 20%. How much does the record actually cost?

8. The cost of a meal is £8 plus value added tax at 15%. How much must I pay for the meal?

9. Miss Kendall earns £120 per week from which income tax is deducted at 30%. Find how much she actually gets. (This is called her *net* pay.)

10. In a certain week a factory worker earns £150 from which income tax is deducted at 30%. Find his net income after tax, i.e. how much he actually gets.

11. Mr Hall earns £1000 per month. If income tax is deducted at 25%, find his net pay after tax.

12. As a result of using Alphamix fertilizer, my potato crop increased by 32% compared with last year. If I grew 150 kg of potatoes last year, what weight of potatoes did I grow this year?

13. The number of children attending Croydly village school is 8% fewer this year than last year. If 450 attended last year, how many are attending this year?

14. The marked price of a man's suit is £125. In a sale the price is reduced by 12%. Find the sale price.

15. In a sale all prices are reduced by 10%. What is the sale price of an article marked   a) £40   b) £85?

16. Last year in Blytham there were 75 reported cases of measles. This year the number of reported cases has dropped by 16%. How many cases have been reported this year?

17. Mr Connah was 115 kg when he decided to go on a diet. He lost 10% of his weight in the first month and a further 8% of his original weight in the second month. How much did he weigh after 2 months of dieting?

18. A car is valued at £8000. It depreciates by 20% in the first year and thereafter each year by 15% of its value at the beginning of that year. Find its value   a) after 2 years   b) after 3 years.

19. In any year the value of a motorcycle depreciates by 10% of its value at the beginning of that year. What is its value after two years if the purchase price was £1800?

20. When John Short increases the speed at which he motors from an average of 40 mph to 50 mph, the number of miles travelled per gallon decreases by 25%. If he travels 36 miles on each gallon when his average speed is 40 mph, how many miles per gallon can he expect at an average speed of 50 mph?

21. When petrol was 50 p per litre I used 700 litres in a year. The price rose by 12% so I reduced my yearly consumption by 12%. Find

a) the new price of a litre of petrol

b) my reduced annual petrol consumption

c) how much more (or less) my petrol bill is for the year.

MIXED EXERCISES

EXERCISE 15c   **1.** Express $\frac{4}{25}$

a) as a percentage

b) as a decimal.

2. Express 0.45

a) as a percentage

b) as a common fraction in its lowest terms.

3. Express 85%

a) as a decimal

b) as a common fraction in its lowest terms.

4. Express 6 mm as a percentage of 3 cm.

5. Find 35% of 120 m².

6. If a number is increased by 25%, what percentage is the new number of the original number?

7. What multiplying factor would increase a quantity by

8. a) Increase 56 cm by 75%.
   b) Decrease 1200 sheep by 20%.

9. The annual cost of insuring the contents of a house is 0.3% of the value of the contents. How much will it cost to insure contents valued at £14 500?

EXERCISE 15d 1. Express $\frac{9}{20}$
   a) as a percentage
   b) as a decimal.

2. Express 0.85
   a) as a percentage
   b) as a common fraction in its lowest terms.

3. Express 64%
   a) as a decimal
   b) as a common fraction in its lowest terms.

4. Express 170 cm as a percentage of 4 m.

5. Find 62% of 3.5 m.

6. If a number is decreased by 42%, what percentage is the new number of the original number?

7. What multiplying factor would decrease a quantity by 18%?

8. a) Increase 70 m by 35%.
   b) Decrease 55 miles by 84%.

9. In a sale a shopkeeper reduces the prices of his goods by 10%. Find the sale price of goods marked   a) £24.50   b) £164.

16 TRIGONOMETRY TANGENT OF AN ANGLE

INVESTIGATING RELATIONSHIPS

In this chapter we are going to look at the relationship between the sizes of the angles and the lengths of the sides in right-angled triangles.

EXERCISE 16a **1.** a) Draw the given triangle accurately using a protractor and a ruler.

b) Measure $\widehat{A}$.

c) Find $\dfrac{BC}{AB}$ as a decimal.

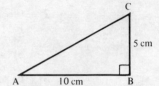

Repeat question 1 for the triangles in questions 2 to 5.

2.
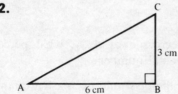

4.
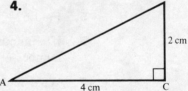

3.
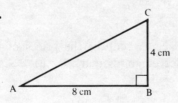

5.
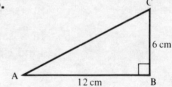

6. Are the triangles in questions 1 to 5 similar?

Repeat question 1 for the triangles in questions 7 to 12.

7.
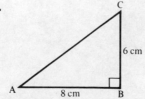

8.
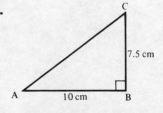

240

9.

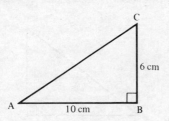

11.

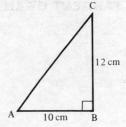

10.

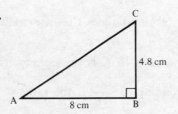

12.

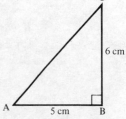

13.

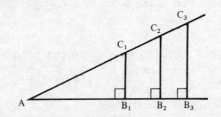

Similar triangles can be drawn so that they overlap as in this diagram. Copy the diagram above on squared paper. Choose your own measurements but make sure that the lengths of the horizontal lines are whole numbers of centimetres. Measure $\hat{A}$.

Find $\dfrac{B_1C_1}{AB_1}$, $\dfrac{B_2C_2}{AB_2}$ and $\dfrac{B_3C_3}{AB_3}$ as decimals.

14. Copy and complete the table using the information from questions 1 to 13.

Angle A	$\dfrac{BC}{AB}$
$26\frac{1}{2}$	0.5

TANGENT OF AN ANGLE

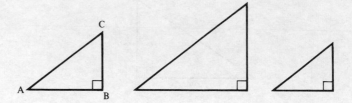

If we consider the set of all triangles that are similar to △ABC then, for every triangle in the set,

the angle corresponding to $\widehat{A}$ is the same

the ratio corresponding to $\dfrac{BC}{AB}$ is the same

where BC is the side *opposite* to $\widehat{A}$

and   AB is the *adjacent* (or neighbouring) side to $\widehat{A}$

From the last exercise you can see that, in a right-angled triangle the ratio $\dfrac{\text{opposite side}}{\text{adjacent side}}$ is always the same for a given angle whatever the size of the triangle.

The ratio $\dfrac{\text{opposite side}}{\text{adjacent side}}$ is called the *tangent* of the angle.

$$\text{tangent of the angle} = \frac{\text{opposite side}}{\text{adjacent side}}$$

Or, briefly,

$$\tan(\text{angle}) = \frac{\text{opp}}{\text{adj}}$$

The information about this ratio is used so often that we need a more complete and more accurate table than the one made in the last exercise. This is given at the end of the book under the heading "Tangents". The information is also stored in some calculators.

FINDING TANGENTS OF ANGLES ———————————————

Using a calculator

To find the tangent of 33°, enter 33 then press the button labelled "tan". You will obtain a number which fills the display. Write down the tangent correct to 4 significant figures.

$$\tan 33° = 0.6494$$

Or, if 3 figures are required, $\tan 33° = 0.649$

Calculators vary; consult your instruction book if you do not get the correct answer.

Using three-figure tables

Notice that the left-hand column gives the angle. The figures in the body of the table give the tangent. Look down the left-hand column for 33°; then in the first column next to it, under the heading .0, you will find the number 0.649.

$$\tan 33° = 0.649$$

EXERCISE 16b

Find the tangent of 56° to 3 s.f.

$$\tan 56° = 1.48$$

Find the tangents of the following angles correct to 3 s.f.:

1.	20°	**7.**	19°	**13.**	4°	**19.**	5°
2.	28°	**8.**	12°	**14.**	37°	**20.**	51°
3.	72°	**9.**	21°	**15.**	44°	**21.**	69°
4.	53°	**10.**	45°	**16.**	89°	**22.**	48°
5.	59°	**11.**	61°	**17.**	52°	**23.**	74°
6.	9°	**12.**	70°	**18.**	35°	**24.**	17°

25. Find the tangents of the angles listed in question 14 in Exercise 16a. How do the answers you now have compare with the decimals you worked out?
If they are different, give a reason for this.

DECIMALS OF DEGREES

Sometimes we need the tangent of an angle which is not a whole number of degrees, for instance 34.2°. To use a calculator, enter 34.2, then press the "tan" key. To use tables, look down the left-hand column for 34, then move across the page until you find the column headed .2 or 0.2. There you will find .680.

$$\tan 34.2° = 0.680$$

(The 0 before the point is printed only in the first column of the tables.)

EXERCISE 16c Find the tangents of the following angles, correct to 3 s.f.:

1.	15.5°	**7.**	30.6°	**13.**	42.4°	**19.**	20.7°
2.	29.6°	**8.**	15.9°	**14.**	71.2°	**20.**	0.7°
3.	11.4°	**9.**	10.2°	**15.**	49.5°	**21.**	70.0°
4.	60.1°	**10.**	3.8°	**16.**	58.8°	**22.**	15.6°
5.	70.7°	**11.**	49.0°	**17.**	65.3°	**23.**	39.9°
6.	46.5°	**12.**	32.7°	**18.**	63.2°	**24.**	44.1°

USING FOUR-FIGURE TABLES

Four-figure tables that use decimals of degrees can be used in a similar way once the use of three-figure tables has been mastered. There is more information stored in the columns on the right of these tables but you can ignore this column for the time being.

To find the tangent of 45.6°, look for 45 down the left-hand column then across under .6 or 0.6.

$$\tan 45.6° = 1.0212 \qquad \text{or} \qquad 1.021$$

Often the figure before the point is printed only in the first column. This is to save space.

EXERCISE 16d Use your calculator or four-figure tables to give the tangents of the following angles correct to 4 s.f.

1.	16.8°	**3.**	81.4°	**5.**	36°	**7.**	58.5°
2.	19°	**4.**	48.9°	**6.**	17.2°	**8.**	62°

THE NAMES OF THE SIDES OF A RIGHT-ANGLED TRIANGLE

Before we can use the tangent for finding sides and angles we need to know which is the side opposite to the given angle and which is the adjacent side.

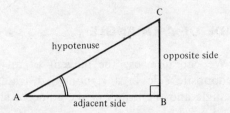

a) The longest side, that is the side opposite the right angle, is called the *hypotenuse*.

b) The side next to the angle (not the hypotenuse) is called the *adjacent side*.

c) The third side is the *opposite side*. It is opposite the particular angle we are concerned with.

Sometimes the triangle is in a position different from the one we have been using.

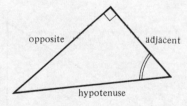

EXERCISE 16e Sketch the following triangles. The angle we are concerned with is marked with a double arc like this ∡. Label the sides "hypotenuse", "adjacent" and "opposite". If necessary, turn the page round so that you can see which side is which.

1. 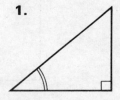   **2.** 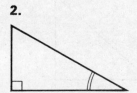   **3.**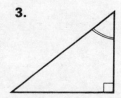

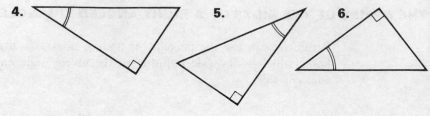

4.   **5.**   **6.**

FINDING A SIDE OF A TRIANGLE

We can now use the tangent of an angle to find the length of the opposite side in a right-angled triangle provided that we know an angle and the length of the adjacent side.

EXERCISE 16f In this exercise, a calculator can be used or tables. Give your answers correct to 3 s.f.

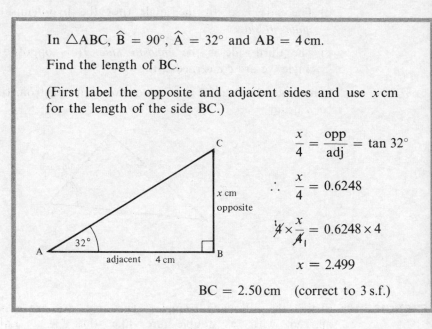

In $\triangle ABC$, $\widehat{B} = 90°$, $\widehat{A} = 32°$ and $AB = 4\,cm$.

Find the length of BC.

(First label the opposite and adjacent sides and use $x\,cm$ for the length of the side BC.)

$$\frac{x}{4} = \frac{opp}{adj} = \tan 32°$$

$$\therefore \quad \frac{x}{4} = 0.6248$$

$$\cancel{4} \times \frac{x}{\cancel{4}} = 0.6248 \times 4$$

$$x = 2.499$$

$$BC = 2.50\,cm \quad \text{(correct to 3 s.f.)}$$

Find the length of BC in questions 1 to 8.

1.

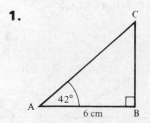

2.

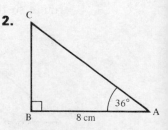

3.

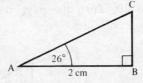

6.

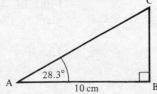

4.

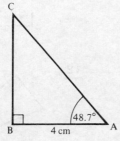

7.

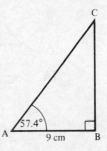

5.

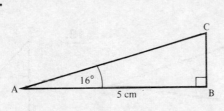

8.

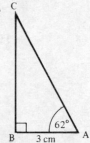

In questions 9 to 12 different letters are used for the vertices of the triangles. In each case find the side required.

9. Find PQ.

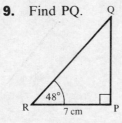

11. Find AC.

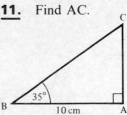

10. Find YZ.

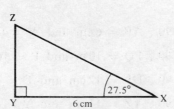

12. Find AC.

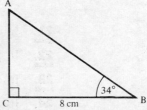

Find BC in questions 13 to 20. Turn the page round if necessary to identify the opposite and adjacent sides.

13.

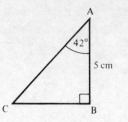

14.

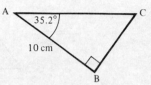

15.

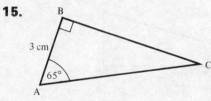

16.

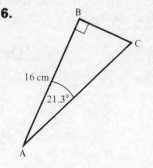

17.

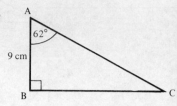

18.

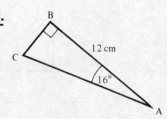

19.

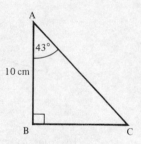

20.

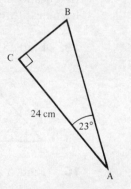

21. In $\triangle ABC$, $\widehat{B} = 90°$, $AB = 6\,cm$ and $\widehat{A} = 41°$. Find BC.

22. In $\triangle PQR$, $\widehat{Q} = 90°$, $PQ = 10\,m$ and $\widehat{P} = 16.7°$. Find QR.

23. In $\triangle DEF$, $\widehat{F} = 90°$, $DF = 12\,cm$ and $\widehat{D} = 56°$. Find EF.

24. In $\triangle XYZ$, $\widehat{Z} = 90°$, $YZ = 11\,cm$ and $\widehat{Y} = 40°$. Find XZ.

FINDING A SIDE ADJACENT TO THE GIVEN ANGLE

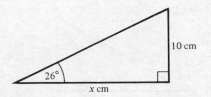

Sometimes the side whose length we are asked to find is adjacent to the given angle instead of opposite to it. Using $\frac{10}{x}$ instead of $\frac{x}{10}$ can lead to an awkward equation so we work out the size of the angle opposite x and use it instead. In this case this other angle is $64°$ and we label the sides "opposite" and "adjacent" to this angle.

Using $64°$,

$$\frac{x}{10} = \frac{\text{opposite}}{\text{adjacent}} = \tan 64°$$

so $\dfrac{x}{10} = 2.05$ giving $x = 20.5$.

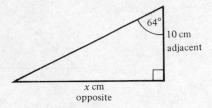

EXERCISE 16g Use a calculator or tables. Give your answers correct to 3 s.f.

In $\triangle PQR$, $\widehat{P} = 90°$, $\widehat{Q} = 51°$ and $PR = 4\,$cm. Find the length of PQ.

(First find the other angle, i.e., $\widehat{R}$.)

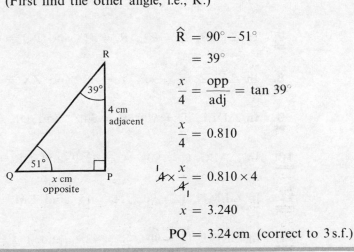

$$\widehat{R} = 90° - 51°$$
$$= 39°$$

$$\frac{x}{4} = \frac{\text{opp}}{\text{adj}} = \tan 39°$$

$$\frac{x}{4} = 0.810$$

$$4 \times \frac{x}{4} = 0.810 \times 4$$

$$x = 3.240$$

$$PQ = 3.24\,\text{cm} \quad \text{(correct to 3 s.f.)}$$

1. Find ZY.

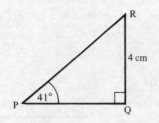

2. Find QP.

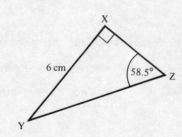

3. Find XZ.

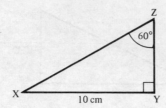

4. Find FD.

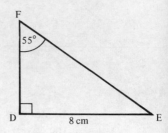

5. Find BC.

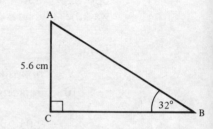

6. Find AB.

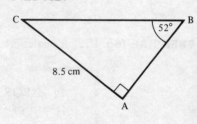

7. In △PQR, $\widehat{Q} = 90°$, $\widehat{R} = 31°$ and PQ = 6 cm. Find RQ.

8. In △XYZ, $\widehat{Z} = 90°$, $\widehat{Y} = 38°$ and ZX = 11 cm. Find YZ.

9. In △DEF, $\widehat{D} = 90°$, $\widehat{E} = 34.8°$ and DF = 24 cm. Find DE.

10. In △ABC, $\widehat{C} = 90°$, $\widehat{A} = 42.4°$ and CB = 3.2 cm. Find AC.

11. In △LMN, $\widehat{L} = 90°$, $\widehat{N} = 15°$ and LM = 4.8 cm. Find LN.

12. In △STU, $\widehat{U} = 90°$, $\widehat{S} = 42.2°$ and TU = 114 cm. Find SU.

PROBLEMS

EXERCISE 16h Give your answers correct to 3 s.f.

> A tree stands on level ground. A is a point on the ground 20 m from the foot of the tree. The angle of elevation of the top, C, from A is 23°. What is the height of the tree?
>
>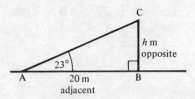
>
> Let BC be h metres
>
> $$\frac{h}{20} = \frac{\text{opp}}{\text{adj}} = \tan 23°$$
>
> $$\frac{h}{20} = 0.424$$
>
> $$20 \times \frac{h}{20} = 0.424 \times 20$$
>
> $$h = 8.480$$
>
> The height of the tree is 8.48 m   (correct to 3 s.f.)

1. In a triangle ABC, $\hat{A}$ is 35°, $\hat{B}$ is 90° and the length of BC is 10 cm. Find the length of AB.

2. Triangle PQR has a right angle at Q, the length of side PQ is 15 cm and $\hat{P}$ is 50°. Find the length of QR.

3. In triangle PQR, $\hat{P}$ is 90° and $\hat{Q}$ is 34.2°. The length of PQ is 12 cm. Find the length of PR.

4. Triangle XYZ has side XY of length 11 cm, $\hat{Y}$ is a right angle and $\hat{X}$ is 42.5°. Find the length of YZ.

5. A pole stands on level ground. A is a point on the ground 10 m from the foot of the pole. The angle of elevation of the top, C, from A is 27°. What is the height of the pole?

6. ABCD is a rectangle. AB = 42 m and BÂC = 59°. Find the length of BC.

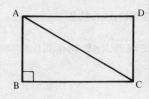

7. In rectangle ABCD, the angle between the diagonal AC and the side AB is 22°, AB = 8 cm. Find the length of BC.

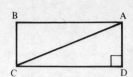

8. In △PQR, PQ = QR. From symmetry, S is the midpoint of PR. P̂ = 72°. PR = 20 cm. Find the height QS of the triangle.

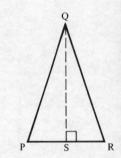

9. A point R is 14 m from the foot of a flagpole PQ. The angle of elevation of the top of the pole from R is 22°. Find the height of the pole.

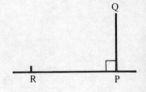

10. A ladder leans against a vertical wall so that it makes an angle of 35° with the wall. The top of the ladder is 2 m up the wall. How far out from the wall is the foot of the ladder?

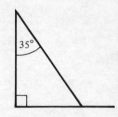

11. A boat B is 60 m out to sea from the foot A of a vertical cliff AC. From C the angle of depression of B is 16°.

a) Find B̂.

b) Find the height of the cliff.

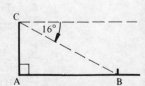

FINDING AN ANGLE GIVEN ITS TANGENT

If we are given the value of the tangent of an angle then we can use a calculator or tables to find that angle.

Using a calculator

To find the angle whose tangent is 0.732, enter 0.732 and then press the inverse button followed by the "tan" button. (If this does not work, consult your instruction book.) The number you see filling the display is the size of the angle in degrees. Give the angle correct to 3 significant figures.

If $\qquad\qquad\qquad\qquad$ $\tan \widehat{A} = 0.732$

then $\qquad\qquad\qquad\qquad$ $\widehat{A} = 36.2°$

Using three-figure tables

Find the number nearest to the tangent value in the body of the table. Then you can see what the angle is by looking at the heading of the column and at the left-hand end of the row. For example, 0.532 is opposite 28° and in the column headed .0

If $\qquad\qquad\qquad\qquad$ $\tan \widehat{A} = 0.532$

$\qquad\qquad\qquad\qquad$ $\widehat{A} = 28.0°$

EXERCISE 16i

> Find the angle whose tangent is 0.516. Give your answer to 3 s.f.
>
> $\qquad\qquad$ $\tan \widehat{A} = 0.516$
>
> $\qquad\qquad$ $\widehat{A} = 27.3°$

Find the angles whose tangents are given below.

1. 2.2	**4.** 4.1	**7.** 0.16
2. 0.36	**5.** 1.4	**8.** 0.62
3. 0.41	**6.** 0.31	**9.** 0.81

If using tables, first correct the given number to 3 d.p. in the following questions. Add zeros if necessary.

10. 0.6752	**13.** 0.56093	**16.** 2.0879
11. 0.99293	**14.** 1.7143	**17.** 2.666666
12. 0.37624	**15.** 3.5843	**18.** 0.33333

19.	0.469	**25.**	0.381	**31.**	1.26
20.	0.256	**26.**	0.574	**32.**	1.1
21.	0.769	**27.**	0.697	**33.**	1.113
22.	0.840	**28.**	0.811	**34.**	1.7
23.	0.975	**29.**	1.14	**35.**	1.01
24.	0.953	**30.**	3.59	**36.**	1.21

EXERCISE 16j Use a calculator or four-figure tables to find the angles whose tangents are given below. Add zeros where necessary if using tables.

1.	0.4245	**7.**	2.056	**13.**	0.3333
2.	0.6847	**8.**	2.4	**14.**	0.74
3.	0.7898	**9.**	1.888	**15.**	1.1263
4.	0.926	**10.**	0.3201	**16.**	1.2218
5.	0.6176	**11.**	0.147	**17.**	1.2366
6.	0.6059	**12.**	0.7357	**18.**	1

TANGENTS IN THE FORM OF FRACTIONS

If we are given the value of a tangent in fraction form, then we need to change it to a decimal before we can find the angle.

EXERCISE 16k

Find the angle whose tangent is $\frac{3}{4}$.

$$\tan \widehat{A} = \frac{3}{4}$$
$$= 0.750$$
$$\widehat{A} = 36.9°$$

Find the angles whose tangents are given below.

1.	$\frac{3}{5}$	**4.**	$\frac{2}{5}$	**7.**	$\frac{5}{4}$	**10.**	$1\frac{1}{2}$
2.	$\frac{4}{5}$	**5.**	$\frac{7}{10}$	**8.**	$\frac{3}{8}$	**11.**	$\frac{3}{25}$
3.	$\frac{1}{2}$	**6.**	$\frac{3}{20}$	**9.**	$2\frac{1}{4}$	**12.**	$2\frac{2}{5}$

Find the angle whose tangent is $\frac{2}{3}$.

$$\tan \widehat{A} = \frac{2}{3}$$
$$= 0.6666\ldots$$
$$= 0.6667 \quad (\text{correct to 4 s.f.})$$
$$\widehat{A} = 33.7°$$

13. $\frac{1}{3}$ **16.** $\frac{5}{6}$ **19.** $\frac{3}{7}$ **22.** $\frac{7}{3}$

14. $\frac{1}{7}$ **17.** $\frac{7}{6}$ **20.** $\frac{2}{9}$ **23.** $\frac{4}{9}$

15. $\frac{1}{6}$ **18.** $\frac{5}{3}$ **21.** $\frac{5}{7}$ **24.** $\frac{4}{3}$

FINDING AN ANGLE GIVEN TWO SIDES OF A TRIANGLE

We can now find an angle in a right-angled triangle if we are given the opposite and adjacent sides.

EXERCISE 16I

In $\triangle ABC$, $\widehat{B} = 90°$, $AB = 8$ cm and $BC = 7$ cm Find $\widehat{A}$.

(First mark the angle and label the opposite and adjacent sides.)

$$\tan \widehat{A} = \frac{\text{opp}}{\text{adj}} = \frac{7}{8}$$
$$= 0.875$$
$$\widehat{A} = 41.2°$$

Find $\widehat{A}$ in questions 1 to 10.

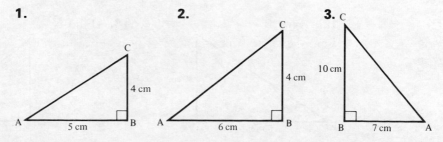

1. A, 5 cm, B, 4 cm, C

2. A, 6 cm, B, 4 cm, C

3. C, 10 cm, B, 7 cm, A

4.

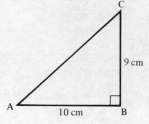

5.

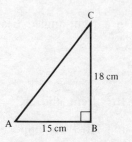

6.

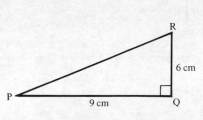

7.

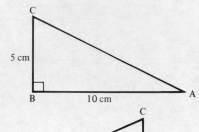

8.

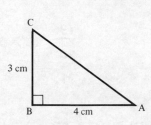

9.

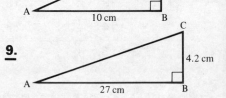

10.

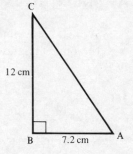

In questions 11 to 16, different letters are used.

11. Find $\hat{P}$.

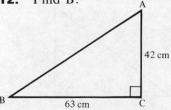

13. Find $\hat{Y}$.

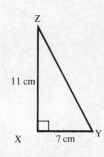

12. Find $\hat{B}$.

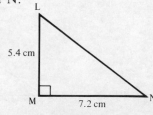

14. Find $\hat{N}$.

15. Find $\hat{D}$.

16. Find $\hat{C}$.

Find $\hat{A}$ in questions 17 to 26. Turn the page round if necessary before labelling the sides.

17.

21.

18.

22.

19.

23.

20.

24.

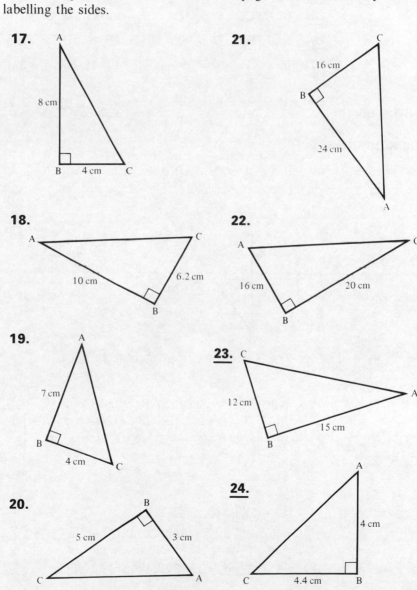

25.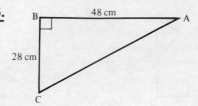

26.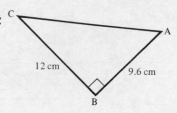

27. In △ABC, $\hat{B} = 90°$, AB = 12 cm, BC = 11 cm. Find $\hat{A}$.

28. In △PQR, $\hat{P} = 90°$, PQ = 3.2 m, PR = 2.8 m. Find $\hat{Q}$.

29. In △DEF, $\hat{D} = 90°$, DE = 108 m, DF = 72 m. Find $\hat{F}$.

30. In △XYZ, $\hat{Z} = 90°$, YZ = 4.5 m, XZ = 3.5 m. Find $\hat{X}$.

PROBLEMS

EXERCISE 16m

A man walks due north for 5 km from A to B, then 4 km due east to C. What is the bearing of C from A?

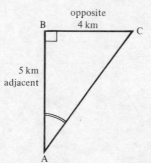

$$\tan \hat{A} = \frac{\text{opp}}{\text{adj}} = \frac{4}{5}$$

$$= 0.800$$

$$\hat{A} = 38.7°$$

The bearing of C from A is 038.7°.

1. ABCD is a rectangle. AB = 60 m and BC = 36 m. Find the angle between the diagonal and the side AB.

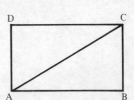

2. A flagpole PQ is 10 m high. R is a point on the ground 20 m from the foot of the pole. Find the angle of elevation of the top of the pole from R (i.e. $\hat{R}$).

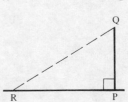

3. In △ABC, AB = BC. AC = 12 cm. D is the midpoint of AC. The height BD of the triangle is 10 cm. Find Ĉ and the other angles of the triangle.

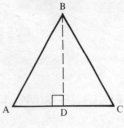

4. A ladder leans against a vertical wall. Its top, Q, is 3 m above the ground and its foot, P, is 2 m from the foot of the wall. Find the angle of slope of the ladder (that is, P̂).

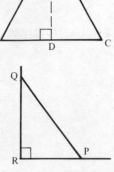

5. The bearing of town A from town B is 032.4°. A is 16 km north of B. How far east of B is it?

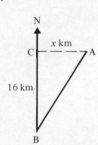

6. In a square, ABCD, of side 8 cm, A is joined to the midpoint E of BC. Find EÂB, CÂB and CÂE. Notice that AE does *not* bisect CÂB.

7. A ladder leans against a vertical wall. It makes an angle of 72° with the horizontal ground and its foot is 1 m from the foot of the wall. How high up the wall does the ladder reach?

8. Sketch axes for x and y from 0 to 5. A is the point $(1, 0)$ and B is $(5, 2)$. What angle does the line AB make with the x-axis?

9. In a rhombus the two diagonals are of lengths 6.2 cm and 8 cm. Find the angles of the rhombus.

10. In rectangle ABCD, AB = 24 cm and BC = 11 cm. Find CÂB and hence find the obtuse angle between the diagonals.

11. In △ABC, AB = BC, CA = 10 cm and Ĉ = 72°. Find the height BD of the triangle.

17 SIMPLE INTEREST

FINDING SIMPLE INTEREST

Everybody wishes to borrow something at one time or another. Perhaps you want to borrow a video camera to record a wedding, a dress to wear to an important event or even a car for a few minutes on a dodgem track. In the same way, the time will come when you will wish to borrow money to buy a motorcycle, a car, furniture or even a house.

The cost of hiring or borrowing money is called the *interest*. The sum of money borrowed (or lent) is called the *principal* and the interest is usually an agreed *percentage* of the sum borrowed.

For example, if £100 is borrowed for a year at an interest rate of 12% per year, then the interest due is $\frac{12}{100}$ of £100, i.e. £12.

The interest due on larger amounts increases in direct proportion, i.e. the interest due on £200 for one year at 12% would be £200 $\times \frac{12}{100}$ = £24, and on £P for one year at 12% would be £P $\times \frac{12}{100}$.

If we double the period of the loan, we double the interest due, and so on. The interest on £P invested for T years at 12% would therefore be £P $\times \frac{12}{100} \times T$.

If the interest rate was R% instead of the given 12%, the interest, I, would be £P $\times \dfrac{R}{100} \times T$.

When interest is calculated this way it is called *simple interest*.

Therefore

$$I = \frac{PRT}{100}$$

where
    I is the simple interest in £s
    P is the principal in £s
    R is the rate per cent per year
    T is the time in years

Unless stated otherwise, R% will always be taken to mean R% each year or per annum.

EXERCISE 17a Find the simple interest on:

1.	£100 for 2 years at 10%	**6.**	£200 for 2 years at 10%
2.	£100 for 2 years at 12%	**7.**	£200 for 5 years at 8%
3.	£100 for 3 years at 8%	**8.**	£300 for 4 years at 12%
4.	£100 for 4 years at 13%	**9.**	£400 for 6 years at 9%
5.	£100 for 7 years at 11%	**10.**	£600 for 7 years at 11%

Find the simple interest on £755 for 6 years at 12%.

$$I = \frac{PRT}{100}$$

$$\therefore \quad \text{simple interest} = £\frac{755 \times 12 \times 6}{100}$$

$$= £543.60$$

The answers to the following questions are exact in £s and pence.

Find the simple interest on:

11.	£350 for 5 years at 7%	**16.**	£484 for 3 years at 7%
12.	£125 for 4 years at 12%	**17.**	£372 for 7 years at 14%
13.	£642 for 7 years at 11%	**18.**	£94 for 6 years at 9%
14.	£1740 for 8 years at 8%	**19.**	£648 for 5 years at 13%
15.	£724 for 3 years at 6%	**20.**	£926 for 9 years at 14%

Find the simple interest on £134.66 for 5 years at 12%, giving your answer correct to the nearest penny.

$$I = \frac{PRT}{100}$$

$$\therefore \quad \text{simple interest} = £\frac{134.66 \times 12 \times 5}{100}$$

$$= £80.796$$

$$= £80.80 \quad \text{correct to the nearest penny}$$

Find, giving your answers correct to the nearest penny, the simple interest on:

21. £526.52 for 2 years at 12%   **26.** £555.45 for 5 years at 9%

22. £94.56 for 4 years at 8%   **27.** £123.72 for 4 years at 8%

23. £142.16 for 5 years at 11%   **28.** £543.89 for 7 years at 9%

24. £813.40 for 4 years at 13%   **29.** £826.92 for 6 years at 7%

25. £627.83 for 3 years at 14%   **30.** £717.47 for 4 years at 17%

Find the simple interest on £276.34 for 7 years at $11\frac{3}{4}\%$, giving your answer correct to the nearest penny.

$$I = \frac{PRT}{100}$$

$$\therefore \text{ simple interest} = £\frac{276.34 \times 11.75 \times 7}{100}$$

$$= £227.289$$

$$= £227.29 \quad \text{correct to the nearest penny}$$

Find, giving your answers correct to the nearest penny, the simple interest on:

31. £154.40 for 4 years at $8\frac{1}{2}\%$ **36.** £73.58 for $5\frac{3}{4}$ years at $9\frac{3}{4}\%$

32. £273.80 for $4\frac{1}{2}$ years at 9%   **37.** £364.88 for $2\frac{3}{4}$ years at $8\frac{1}{4}\%$

33. £527.49 for 3 years at $12\frac{3}{4}\%$ **38.** £2034.48 for $1\frac{1}{2}$ years at $7\frac{1}{5}\%$

34. £436.15 for $7\frac{1}{2}$ years at $11\frac{1}{4}\%$ **39.** £613.27 for $3\frac{1}{4}$ years at $15\frac{1}{2}\%$

35. £84.72 for $4\frac{1}{4}$ years at $13\frac{1}{2}\%$ **40.** £454.92 for $6\frac{1}{4}$ years at $18\frac{1}{4}\%$

Find the simple interest on £286 for 240 days at 9%.

Since the time, T, must be in years, $T = \dfrac{240}{365}$

$$I = \frac{PRT}{100}$$

$$\therefore \text{ simple interest} = £\frac{286 \times 9}{100} \times \frac{240}{365}$$

$$= £16.924$$

$$= £16.92 \quad \text{correct to the nearest penny}$$

Find, giving your answers correct to the nearest penny, the simple interest on:

41. £320 for 100 days at 12%   **44.** £282.50 for 214 days at 16%

42. £413 for 150 days at 8%   **45.** £613.94 for 98 days at $14\frac{1}{2}$%

43. £1000 for 300 days at 9%   **46.** £729.32 for 22 days at 11%

AMOUNT

If I borrow £250 for 3 years at 11% simple interest, the *sum* of the *interest* and *principal* is the total I must repay to clear the debt. This sum is called the *amount* and is denoted by *A*,

i.e.

$$A = P + I$$

In this case

$$I = £\frac{250 \times 11 \times 3}{100}$$

$$= £82.50$$

$$\therefore \quad \text{amount} = £250 + 82.50$$

$$= £332.50$$

EXERCISE 17b Find the amount of:

1. £350 for 5 years at 10%   **5.** £820 for 8 years at 14%

2. £420 for 2 years at 8%   **6.** £970 for 7 years at 9%

3. £650 for 4 years at 12%   **7.** £492 for 5 years at $8\frac{1}{2}$%

4. £513 for 4 years at $13\frac{1}{2}$%   **8.** £654.20 for 4 years at 9%

9. £738 for $3\frac{1}{2}$ years at 9%   **13.** £192.63 for 5 years at 11%

10. £186 for $4\frac{1}{4}$ years at 12%   **14.** £564.27 for $6\frac{1}{2}$ years at 12%

11. £285 for 9 years at 6%   **15.** £718.55 for $4\frac{1}{4}$ years at $13\frac{1}{2}$%

12. £826.50 for 6 years at 8%   **16.** £318 for $5\frac{3}{4}$ years at $11\frac{1}{2}$%

INVERSE QUESTIONS ON SIMPLE INTEREST

Consider the question of finding the rate of interest if the cost of borrowing £500 for 3 years is £180.

Method 1:   If we substitute these values in the formula $I = \dfrac{PRT}{100}$

we have $180 = \dfrac{\cancel{500}^{5} \times R \times 3}{\cancel{100}_{1}}$ where R is the rate %,

i.e. $180 = 15R$

$$\therefore \quad R = \frac{180}{15} = 12$$

i.e. the rate of interest is 12%.

Method 2:   Make R the subject of the formula $I = \dfrac{PRT}{100}$

i.e. $PRT = 100I$ (multiplying each side by 100)

$$\therefore \quad R = \frac{100I}{PT} \qquad \text{(dividing each side by } PT)$$

Then   rate per cent $= \dfrac{100 \times 180}{500 \times 3}$

$$= 12$$

i.e. the rate of interest is 12%

Similar reasoning is required if either P or T is the unknown quantity.

EXERCISE 17c

Find the principal that will earn £176 in 4 years at 8% simple interest.

Using $I = \dfrac{PRT}{100}$

$$176 = \frac{P \times 8 \times 4}{100}$$

i.e. $17\,600 = 32P$

$$\therefore \quad P = \frac{17\,600}{32}$$

$$= 550$$

The principal is therefore £550.

Find the principal that will earn:

1. £200 simple interest in 5 years at 8%

2. £432 simple interest in 6 years at 12%

3. £196 simple interest in 8 years at 7%

4. £80.85 simple interest in $3\frac{1}{2}$ years at $5\frac{1}{2}$%

5. £396.90 simple interest in $4\frac{1}{2}$ years at $10\frac{1}{2}$%

What is the rate per cent if the cost of borrowing:

6. £500 for 5 years is £250

7. £450 for 4 years is £270

8. £700 for 8 years is £672

9. £850 for 3 years is £204

10. £340 for 6 years is £142.80

What is the rate per cent if:

11. £250 will earn £75 simple interest in 3 years

12. £370 will earn £296 simple interest in 5 years

13. £640 will earn £204.80 simple interest in 4 years

14. £870 will earn £626.40 simple interest in 6 years

15. £435 will earn £156.60 simple interest in 4 years

Find the number of years in which £140 will earn £29.40 at 7%.

Using
$$I = \frac{PRT}{100}$$

$$29.40 = \frac{140 \times 7 \times T}{100}$$

i.e.
$$2940 = 140 \times 7 \times T$$

$$\therefore \quad T = \frac{2940}{140 \times 7}$$

$$= 3$$

The investment is therefore for 3 years.

Find the number of years in which:

16. £300 invested at 8% simple interest will earn £48

17. £535 invested at 12% simple interest will earn £321

18. £470 invested at 14% simple interest will earn £197.40

19. £617 invested at 9% simple interest will earn £222.12

20. £824 invested at 11% simple interest will earn £271.92

What sum of money will amount to £489.60 if invested for 3 years at 12% simple interest?

$$I = \frac{PRT}{100}$$

$$= \frac{P \times 12 \times 3}{100}$$

But $$A = P + I$$

$$\therefore \quad 489.60 = P + \frac{36}{100}P$$

$$= \frac{136}{100}P$$

i.e. $$48\,960 = 136P$$

or $$P = \frac{48\,960}{136}$$

$$= 360$$

Therefore £360 will amount to £489.60 if invested for 3 years at 12% simple interest.

What sum of money will amount to:

21. £348 if invested for 2 years at 8%

22. £841 if invested for 3 years at 15%

23. £688 if invested for 5 years at 12%

24. £1249.20 if invested for 7 years at $10\frac{1}{2}$%

25. £1427.50 if invested for $4\frac{1}{2}$ years at $9\frac{1}{2}$%

PROBLEMS

EXERCISE 17d Copy and complete the following table:

	Principal in £s	Rate %	Time	Simple interest in £s	Amount in £s
1.	230	10	2 years		
2.	180		8 years	230.40	
3.	950	9			1292
4.		18		752.40	1512.40
5.	637		6 years		1210.30
6.		14		44.52	468.52
7.	828		5 months	$25.87\frac{1}{2}$	
8.	555	$12\frac{1}{2}$	144 days		

9. Mr. Sadler's bank pays interest at 8% p.a. on money he has on deposit. How much is in his account if the interest for 7 months is £25.06?

10. Miss Zeraschi pays £84.87 when she borrows a sum of money from the bank for 9 months at 12% p.a. simple interest. How much does she borrow?

11. The interest I receive on £1500 invested for a certain time would increase by £225 if the interest rate rose by 3%. For how long is the sum invested?

12. The interest I receive on £846 would decrease by £118.44 if the interest rate dropped by 2%. For how long is the sum invested?

13. A factory owner borrows £18 600 from his bank to buy machinery. If the annual rate of interest is $12\frac{1}{2}$%, how much must he repay the bank after 9 months to clear the debt?

14. Jane Peters borrowed £219 at 14% p.a. When she repaid the debt the interest due was £10.50. For how many days did she borrow the money?

15. Find the sum to which £1716 will amount in 10 months if it is invested in the Sunly Building Society which pays $8\frac{1}{2}$% per annum.

MIXED EXERCISES

EXERCISE 17e
1. Find the simple interest on £682 invested for 3 years at 8%.

2. Find the simple interest on £85 invested for $2\frac{1}{2}$ years at 5%.

3. What sum of money will give £79.80 simple interest if it is invested for 3 years at 7%?

4. Find the simple interest on £182.73 invested for $2\frac{1}{2}$ years at 4%, giving your answer correct to the nearest penny.

5. Find the amount of £950 for 5 years at 14%.

6. Find the simple interest on £650 for 100 days at 12%.

7. For how many years must £60 be invested at 9% simple interest to give £27?

8. What rate per cent would give £49 simple interest on £350 invested for $3\frac{1}{2}$ years?

9. What rate per cent would give £10 simple interest on £600 invested for 4 months?

10. What sum of money will amount to £860 if invested for 6 years at 12% simple interest?

EXERCISE 17f
1. Find the simple interest on £537 for 5 years at 11%.

2. Find the simple interest on £92 for $3\frac{3}{4}$ years at 9%.

3. Find the sum to which £525 will amount if invested for $6\frac{1}{2}$ years at 12% simple interest.

4. For how many years must £95 be invested at $7\frac{1}{2}$% simple interest to give £28.50?

5. Find the simple interest on £174.54 for 4 years at 14%, giving your answer correct to the nearest penny.

6. If the simple interest on £375 in 3 years is £168.75, find the annual rate of interest.

7. Find the amount of £737 for 8 years at 12%.

8. Find the simple interest on £255.50 for 320 days at 8%.

9. What sum of money will amount to £913.25 if invested at 9% simple interest for $4\frac{1}{2}$ years?

10. The simple interest I received on a sum of money which had been invested for 1 year at 16% decreased by £25.50 the following year when the annual rate fell by 3%. Find the sum invested.

18 VOLUMES CONSTANT CROSS-SECTION

VOLUME OF A CUBOID

Reminder: We find the volume of a cuboid (that is, a rectangular block) by multiplying length by width by height,

i.e.   volume = length × width × height

or $V = l \times w \times h$

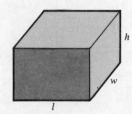

Remember that the measurements must all be in the same units before they are multiplied together.

EXERCISE 18a

Find the volume of a cuboid of length 10 cm, width 66 mm and height 7 cm.

$$\text{Width} = 66\,\text{mm} = 6.6\,\text{cm}$$

$$V = l \times w \times h$$

$$= 10 \times 6.6 \times 7\,\text{cm}^3$$

$$\text{Volume} = 462\,\text{cm}^3$$

1. Find the volume of a cuboid of length 9 cm, width 6 cm and height 4 cm.

2. Find the volume of a cuboid of length 12 m, width 8 m and height 4.5 m.

3. Find the volume of a cuboid of length 300 cm, width 20 cm and height 30 cm.

4. Find the height of a cuboid of length 6.2 cm, width 3.4 cm and height 5 cm.

Find the volumes of the following cuboids, changing the units first if necessary. Do *not* draw a diagram.

	Length	Width	Height	Volume units
5.	3.2 cm	5 mm	10 mm	mm³
6.	$3\frac{1}{4}$ cm	4 cm	$4\frac{1}{2}$ cm	cm³
7.	1.4 cm	9 mm	3.2 mm	mm³
8.	9.2 m	300 cm	1.8 m	m³
9.	0.02 cm	0.04 cm	0.01 cm	cm³
10.	6.2 m	32 mm	20 cm	cm³
11.	$7\frac{1}{2}$ cm	$2\frac{1}{2}$ cm	6 cm	cm³
12.	4.2 cm	3 cm	0.15 m	cm³
13.	7.2 cm	3.6 cm	5 cm	cm³
14.	5.6 m	7 m	3.4 m	m³
15.	7.23 cm	50 mm	4 cm	cm³
16.	4.8 cm	3.2 m	1.5 cm	cm³

VOLUMES OF SOLIDS WITH UNIFORM CROSS-SECTIONS

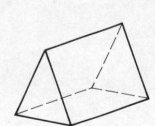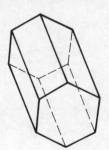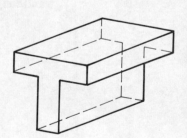

When we cut through any one of the solids above, parallel to the ends, we always get the same shape as the end. This shape is called the cross-section.

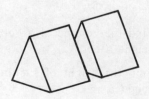

As the cross-section is the same shape and size wherever the solid is cut, the cross-section is said to be *uniform* or *constant*. These solids are also called *prisms* and we can find the volumes of some of them.

First consider a cuboid (which can also be thought of as a rectangular prism).

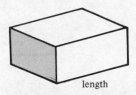

Volume = length × width × height

= (width × height) × length

= area of shaded end × length

= area of cross-section × length

Now consider a triangular prism. If we enclose it in a cuboid we can see that its volume is half the volume of the cuboid.

Volume = ($\frac{1}{2}$ × width × height) × length

= area of shaded triangle × length

= area of cross-section × length

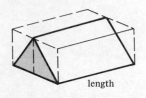

This is true of any prism so that

Volume of a prism = area of cross-section × length

EXERCISE 18b

Find the volume of the solid below.

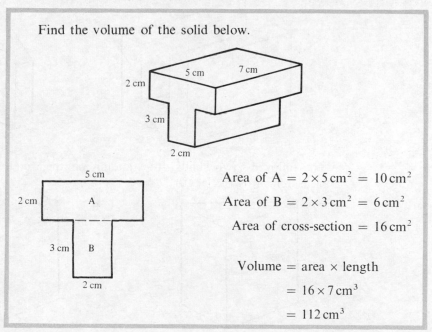

Area of A = $2 \times 5 \, \text{cm}^2 = 10 \, \text{cm}^2$

Area of B = $2 \times 3 \, \text{cm}^2 = 6 \, \text{cm}^2$

Area of cross-section = $16 \, \text{cm}^2$

Volume = area × length

= $16 \times 7 \, \text{cm}^3$

= $112 \, \text{cm}^3$

Find the volumes of the following prisms. Draw a diagram of the cross-section but do *not* draw a picture of the solid.

1.

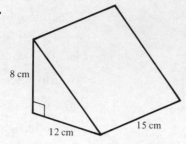

8 cm

12 cm   15 cm

4.

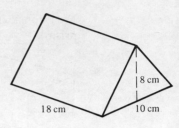

18 cm   8 cm   10 cm

2.

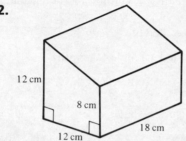

12 cm

8 cm   18 cm

12 cm

5.

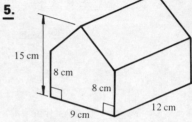

15 cm   8 cm

8 cm

9 cm   12 cm

3.

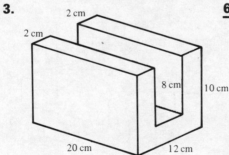

2 cm

2 cm

8 cm   10 cm

20 cm   12 cm

6.

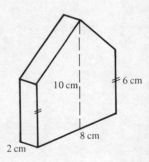

10 cm   6 cm

8 cm

2 cm

The following two solids are standing on their ends so the vertical measurement is the length.

7.

15 cm

11 cm   8 cm

8.

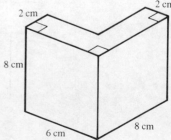

2 cm   2 cm

8 cm

6 cm   8 cm

In questions 9 to 16, the cross-sections of the prisms and their lengths are given. Find their volumes.

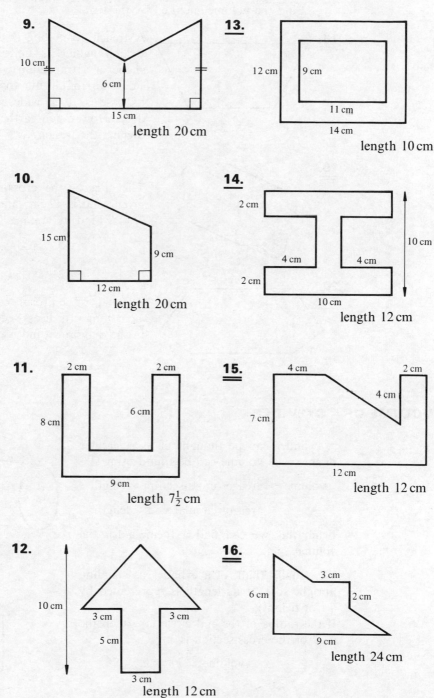

9.

10 cm

6 cm

15 cm

length 20 cm

10.

15 cm

9 cm

12 cm

length 20 cm

11.

2 cm   2 cm

8 cm   6 cm

9 cm

length 7½ cm

12.

10 cm

3 cm   3 cm

5 cm

3 cm

length 12 cm

13.

12 cm   9 cm

11 cm

14 cm

length 10 cm

14.

2 cm

4 cm   4 cm

2 cm

10 cm

10 cm

length 12 cm

15.

4 cm   2 cm

4 cm

7 cm

12 cm

length 12 cm

16.

3 cm

6 cm   2 cm

9 cm

length 24 cm

17. A tent is in the shape of a triangular prism. Its length is 2.4 m, its height 1.8 m and the width of the triangular end is 2.4 m. Find the volume enclosed by the tent.

18.

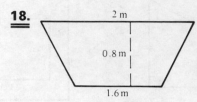

A trench 15 m long is dug. Its cross-section, which is uniform, is in the shape of a trapezium with its parallel sides horizontal. Its top is 2 m wide, its base is 1.6 m wide and it is 0.8 m deep. How much earth is removed in digging the trench?

19.

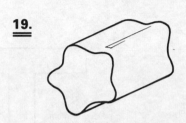

The area of the cross-section of the given solid is 42 cm² and the length is 32 cm. Find its volume.

20.

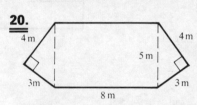

A solid of uniform cross-section is 12 m long. Its cross-section is shown in the diagram. Find its volume.

VOLUME OF A CYLINDER

A cylinder can be thought of as a circular prism so its volume can be found using

volume = area of cross-section × length

= area of circular end × length

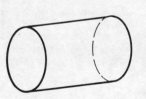

From this we can find a formula for the volume.

We usually think of a cylinder as standing upright so that its length is represented by h (for height).

If the radius of the end circle is r, then the area of the cross-section is πr^2

$$\therefore \quad \text{volume} = \pi r^2 \times h$$

$$= \pi r^2 h$$

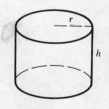

EXERCISE 18c

Find the volume of a cylinder of radius 4 cm and height 6 cm. Use $\pi \approx 3.142$.

Method 1:

Area of cross-section $= \pi r^2$

$= (3.142 \times 4 \times 4) \text{ cm}^2$

$= 50.27 \text{ cm}^2$

Volume $=$ area of cross-section $\times$ length

$= (50.27 \times 6) \text{ cm}^3$

$= 301.62 \text{ cm}^3$

$= 302 \text{ cm}^3$ (to 3 s.f.)

Method 2:   Volume $= \pi r^2 h$

$= (3.142 \times 4 \times 4 \times 6) \text{ cm}^3$

$= 301.62 \text{ cm}^3$

$= 302 \text{ cm}^3$ (to 3 s.f.)

Use $\pi \approx 3.142$ and give all your answers correct to 3 s.f.

Find the volumes of the following cylinders:

1. Radius 2 cm, height 10 cm

6. Radius 1 cm, height 4.8 cm

2. Radius 3 cm, height 4 cm

7. Diameter 4 cm, height 3 cm

3. Radius 5 cm, height 4 cm

8. Diameter 6 cm, height 1.8 cm

4. Radius 3 cm, height 2.1 cm

9. Radius 12 cm, height 10 cm

5. Diameter 2 cm, height 1 cm

10. Radius 7 cm, height 9 cm

11. Radius 3.2 cm, height 10 cm

16. Diameter 2.4 cm, height 6.2 cm

12. Radius 6 cm, height 3.6 cm

17. Radius 4.8 mm, height 13 mm

13. Diameter 10 cm, height 4.2 cm

18. Diameter 16.2 cm, height 4 cm

14. Radius 7.2 cm, height 4 cm

19. Radius 76 cm, height 88 cm

15. Diameter 64 cm, height 22 cm

20. Diameter 0.02 m, height 0.14 m

COMPOUND SHAPES

EXERCISE 18d Find the volumes of the following solids. Take $\pi \approx 3.142$ and give your answers correct to 3 s.f. Draw diagrams of the cross-sections but do *not* draw pictures of the solids.

1. 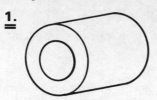 A tube of length 20 cm. The inner radius is 3 cm and the outer radius is 5 cm.

2. 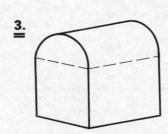 A half-cylinder of length 16 cm and radius 4 cm.

3. 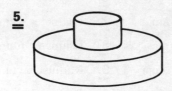 A solid of length 6.2 cm, whose cross-section consists of a square of side 2 cm surmounted by a semicircle.

4.  A disc of radius 9 cm and thickness 0.8 cm.

5. 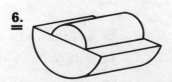 A solid made of two cylinders each of height 5 cm. The radius of the smaller one is 2 cm and of the larger one is 6 cm.

6.  A solid made of two half-cylinders each of length 11 cm. The radius of the larger one is 10 cm and the radius of the smaller one is 5 cm.

19 SINE AND COSINE OF AN ANGLE

TRIGONOMETRY: SINE OF AN ANGLE

The tangent of an angle was useful when the opposite and adjacent sides of a right-angled triangle were involved.

Sometimes we are interested, instead, in the opposite side and the hypotenuse. These two sides form a different ratio which is called the sine of the angle (or sin for short).

In this diagram

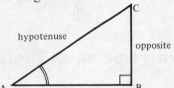

$$\sin \widehat{A} = \frac{opp}{hyp} = \frac{CB}{AC}$$

All right-angled triangles containing an angle of 40°, for example, are similar so $\dfrac{CB}{AC}$ always has the same value.

In all right-angled triangles,

$$\sin \widehat{A} = \frac{opp}{hyp}$$

The values of this ratio for all acute angles are given in the sine tables at the back of the book and also stored in most calculators. We use sine tables in the same way as we use tangent tables.

EXERCISE 19a

> Find the sine of   a) 72°   b) 32.8°
>
> a)   sin 72° = 0.951
> b)   sin 32.8° = 0.542

Find the sines of the following angles:

1.	26°	**6.**	72°
2.	84°	**7.**	16.8°
3.	25.4°	**8.**	4.2°
4.	37.1°	**9.**	62.4°
5.	78.9°	**10.**	71.1°

Find the angle whose sine is 0.909

$$\sin \widehat{A} = 0.909$$
$$\widehat{A} = 65.4°$$

Find the angles whose sines are given below.

11. 0.834   **16.** 0.07

12. 0.413   **17.** 0.647

13. 0.639   **18.** 0.357

14. 0.704   **19.** 0.428

15. 0.937   **20.** 0.261

USING THE SINE RATIO TO FIND A SIDE OR AN ANGLE

EXERCISE 19b

In $\triangle ABC$, $\widehat{B} = 90°$, $\widehat{A} = 28°$ and $AC = 7$ cm.
Find the length of BC.

(Label the sides first.)

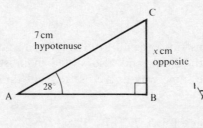

$$\frac{x}{7} = \frac{\text{opp}}{\text{hyp}} = \sin 28°$$

$$\frac{x}{7} = 0.469$$

$$\not7 \times \frac{x}{\not7} = 0.469 \times 7$$

$$x = 3.283$$

$$\therefore \quad BC = 3.28 \text{ cm} \quad \text{(to 3 s.f.)}$$

1. Find BC.

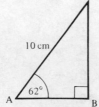

2. Find BC.

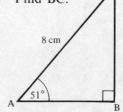

3. Find AC.

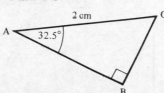

7. Find PQ.

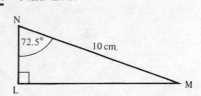

4. Find BC.

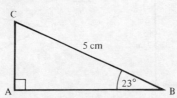

8. Find XY.

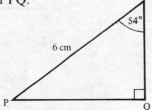

5. Find QR.

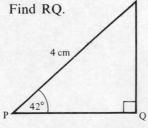

9. Find LM.

6. Find RQ.

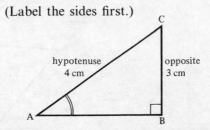

10. Find PQ.

In $\triangle ABC$, $\widehat{B} = 90°$. $AC = 4$ cm and $BC = 3$ cm. Find A.

(Label the sides first.)

$$\sin \widehat{A} = \frac{\text{opp}}{\text{hyp}} = \frac{3}{4}$$

$$= 0.75$$

$$\widehat{A} = 48.6°$$

11. Find $\widehat{A}$.

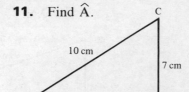

12. Find $\widehat{A}$.

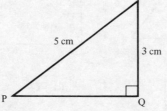

13. Find $\widehat{P}$.

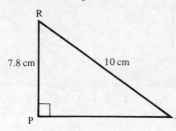

14. Find $\widehat{Q}$.

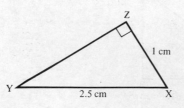

15. Find $\widehat{Y}$.

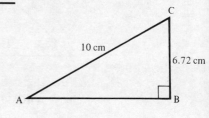

16. Find $\widehat{P}$.

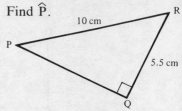

17. Find $\widehat{X}$.

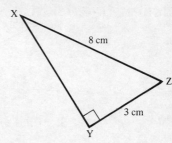

18. Find $\widehat{M}$.

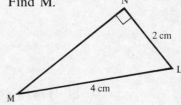

19. Find $\widehat{A}$.

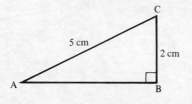

20. Find $\widehat{Q}$.

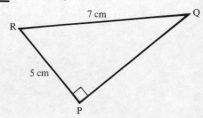

21. In $\triangle ABC$, $\hat{B} = 90°$, $\hat{C} = 36°$ and $AC = 3.5$ cm. Find AB.

22. In $\triangle PQR$, $\hat{R} = 90°$, $PQ = 7$ cm and $\hat{P} = 71.6°$. Find QR.

23. In $\triangle ABC$, $\hat{B} = 90°$, $AB = 3.2$ cm and $AC = 4$ cm. Find $\hat{C}$ and $\hat{A}$.

24. In $\triangle PQR$, $\hat{Q} = 90°$, $PQ = 2.6$ cm and $PR = 5.5$ cm. Find $\hat{R}$.

25. In $\triangle XYZ$, $\hat{X} = 90°$, $YZ = 4.2$ cm and $\hat{Z} = 62.4°$. Find XY.

COSINE OF AN ANGLE

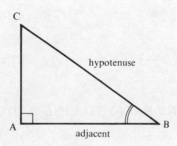

If we are given the adjacent side and the hypotenuse, then we can use a third ratio, $\dfrac{\text{adjacent side}}{\text{hypotenuse}}$. This is called the *cosine* of the angle (cos for short).

$$\cos \hat{B} = \frac{\text{adj}}{\text{hyp}} = \frac{AB}{BC}$$

Cosines of acute angles are given in a table at the back of this book and also stored in most calculators.

EXERCISE 19c

Find the cosine of   a) 41°   b) 28.7°

    a) $\cos 41° = 0.755$

    b) $\cos 28.7° = 0.877$

Find the cosines of the following angles:

1. 59°	**5.** 60.1°	**9.** 17.5°
2. 48°	**6.** 67°	**10.** 25.3°
3. 4°	**7.** 82°	**11.** 86°
4. 44.9°	**8.** 13.8°	**12.** 10°

Find the angle whose cosine is 0.493

$$\cos \widehat{A} = 0.493$$
$$\widehat{A} = 60.5°$$

In questions 13 to 27, $\cos \widehat{A}$ is given. Find $\widehat{A}$.

13.	0.435	**18.**	0.943	**23.**	0.012
14.	0.909	**19.**	0.820	**24.**	0.739
15.	0.714	**20.**	0.567	**25.**	0.628
16.	0.7	**21.**	0.24	**26.**	0.143
17.	0.254	**22.**	0.938	**27.**	0.843

USING THE COSINE RATIO TO FIND A SIDE OR AN ANGLE

EXERCISE 19d

In $\triangle ABC$, $\widehat{B} = 90°$ and $AC = 9\,\text{cm}$.

Find AB.

(Label the sides first.)

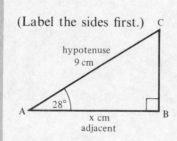

$$\frac{x}{9} = \frac{\text{adj}}{\text{hyp}} = \cos 28°$$

$$\frac{x}{9} = 0.883$$

$$\cancel{9} \times \frac{x}{\cancel{9}} = 0.883 \times 9$$

$$x = 7.947$$

$$AB = 7.95\,\text{cm} \quad (\text{correct to } 3\,\text{s.f.})$$

In the following triangles find the required lengths.

1. Find AB.

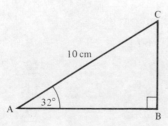

2. Find AB.

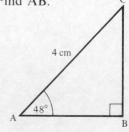

3. Find PQ.

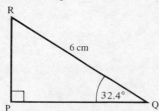

7. Find AC.

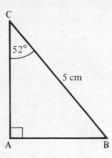

4. Find PQ.

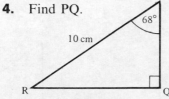

8. Find XZ.

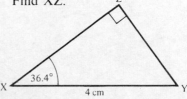

5. Find PQ.

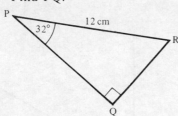

9. Find PR.

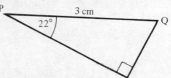

10. Find BC.

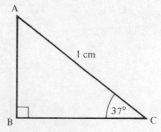

6. Find AB.

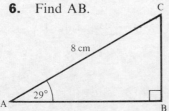

In $\triangle ABC$, $\widehat{B} = 90°$, $AB = 4\,cm$ and $AC = 6\,cm$.
Find $\widehat{A}$.

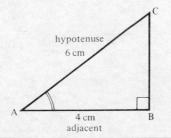

$$\cos \widehat{A} = \frac{adj}{hyp} = \frac{4}{6}$$

$$= 0.667$$

$$\widehat{A} = 48.2°$$

11. Find $\widehat{A}$.

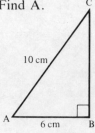

12. Find $\widehat{A}$.

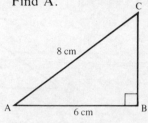

13. Find $\widehat{Y}$.

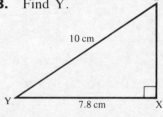

14. Find $\widehat{C}$.

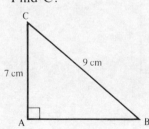

15. Find $\widehat{R}$.

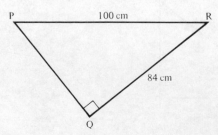

16. Find $\widehat{A}$.

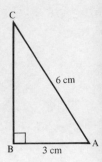

17. Find $\widehat{Q}$.

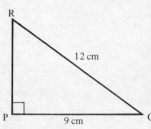

18. Find $\widehat{C}$.

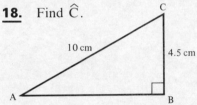

19. Find $\widehat{P}$.

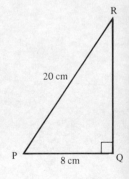

20. Find $\widehat{X}$.

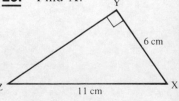

USE OF ALL THREE RATIOS

To remember which ratio is called by which name, some people use the word SOHCAHTOA.

$$\text{Sin } \widehat{A} = \frac{\text{Opposite}}{\text{Hypotenuse}} \quad \text{SOH}$$

$$\text{Cos } \widehat{A} = \frac{\text{Adjacent}}{\text{Hypotenuse}} \quad \text{CAH}$$

$$\text{Tan } \widehat{A} = \frac{\text{Opposite}}{\text{Adjacent}} \quad \text{TOA}$$

EXERCISE 19e

State whether sine, cosine or tangent should be used for the calculation of the marked angle.

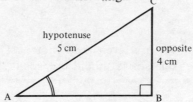

The opposite side and the hypotenuse are given so we should use sin $\widehat{A}$.

In questions 1 to 6 label the sides whose lengths are known, "hypotenuse", "opposite" or "adjacent". Then state whether sine, cosine or tangent should be used for the calculation of the marked angle.

1.

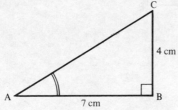

3.

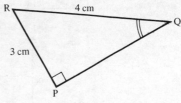

2.

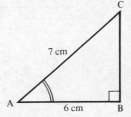

4.

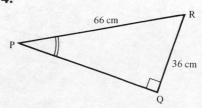

5.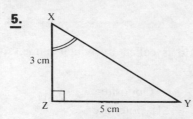

6.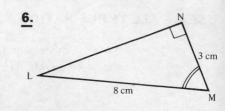

State whether sine, cosine or tangent should be used for the calculation to find x.

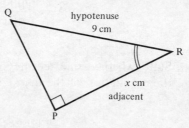

We are given the hypotenuse and wish to find the adjacent side so we should use $\cos \hat{R}$.

In questions 7 to 12, using "opposite", "adjacent" or "hypotenuse", label the side whose length is given and the side whose length is to be found. Then state whether sine, cosine or tangent should be used for the calculation to find x.

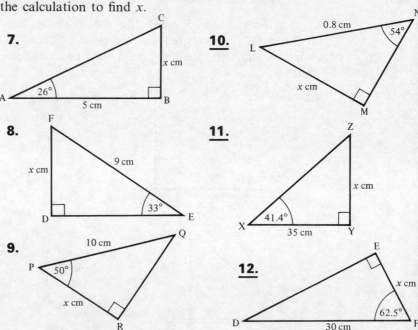

7.

10.

8.

11.

9.

12.

13. Calculate the marked angles in questions 1 to 6 and the lengths given as *x* cm in questions 7 to 12.

In questions 14 to 19, find the marked angles.

14.

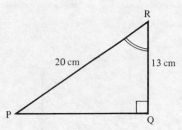

17.

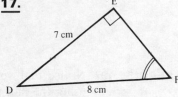

15.

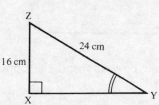

18.

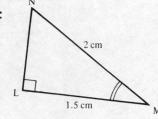

16.

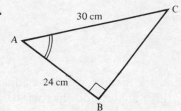

19.

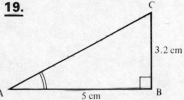

In questions 20 to 25, find the length of the side marked *x* cm.

20.

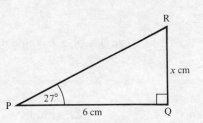

22.

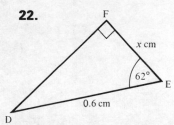

21.

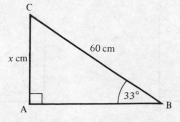

23.

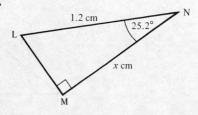

24.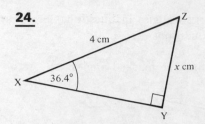

25.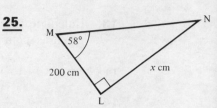

EXERCISE 19f Notice that you can find the size of the third angle of the triangle by using the fact that the three angles add up to 180°.

In △ABC, $\widehat{B} = 90°$, AC = 15 cm and BC = 11 cm. Find $\widehat{A}$, then $\widehat{C}$.

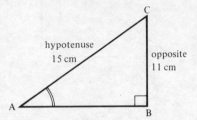

(We are given the hypotenuse and the opposite side so we should use $\sin \widehat{A}$.)

$$\sin \widehat{A} = \frac{\text{opp}}{\text{hyp}} = \frac{11}{15}$$

$$= 0.7333$$

$$\widehat{A} = 47.2°$$

$$\widehat{C} = 90° - 47.2°   \qquad \text{(angles of a triangle add up to 180°)}$$

$$= 42.8°$$

1. Find $\widehat{A}$, then $\widehat{C}$.

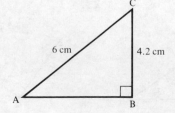

2. Find AC.

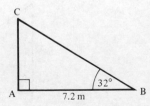

3. Find $\hat{X}$, then $\hat{Z}$.

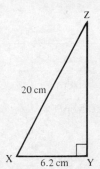

4. Find LM.

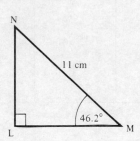

5. Find PQ.

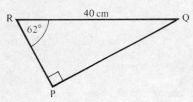

6. Find $\hat{C}$, then AB.

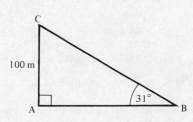

7. Find AB.

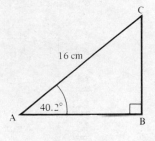

8. Find PQ.

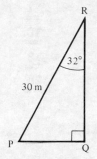

9. Find XZ.

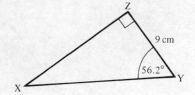

10. Find $\hat{A}$.

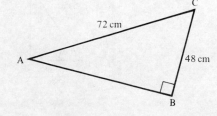

In questions 11 to 20 ABC is a triangle in which $\widehat{B} = 90°$. Find the length or angle marked with a cross.

	AB	BC	CA	$\widehat{A}$	$\widehat{C}$
11.	7 cm		10 cm	×	
12.		×	5 cm	32.6°	
13.	6 cm	×		18°	×
14.		2.4 cm	6 cm		×
15.	7 cm	9 cm		×	×
16.		×	8 cm		46°
17.		2.42 cm	4 cm		×
18.	20 cm		35 cm	×	
19.	16 cm	22 cm		×	×
20.	×	×	20 cm	32°	

PROBLEMS

EXERCISE 19g

In an isosceles triangle PQR, $PQ = QR = 5$ cm and $PR = 6$ cm. Find the angles of the triangle.

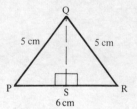

(Divide the triangle down the middle.)

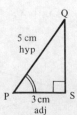

$$\cos \widehat{P} = \frac{\text{adj}}{\text{hyp}} = \frac{3}{5}$$

$\widehat{P} = 53.1°$

$\widehat{R} = 53.1°$ (isosceles $\triangle$; base angles equal)

$P\widehat{Q}R = 73.8°$ (angles of a $\triangle$ add up to 180°)

1. In rectangle ABCD, AC = 4 cm and BC = 3 cm.
Find CÂB.

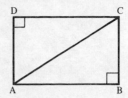

2. C is the centre of a circle of radius 10 cm. CÂB = 31°.
Find the distance of the chord AB from the centre, i.e. find DC.

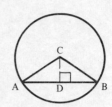

3. C is the centre of a circle of radius 4 cm. Chord AB is of length 5 cm.
Find CÂB.

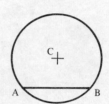

4. A ladder 2 m long leans against a wall. Its top is 1.6 m above the foot of the wall.
Find the angle that the ladder makes with the ground.

5. A ladder 4 m long stands on horizontal ground and leans against a vertical wall. It makes an angle of 25° with the wall. How far is the foot of the ladder from the foot of the wall?

6. In △ABC, AB = AC = 8 cm and B̂ = 68.6°.
Find the height of the triangle.

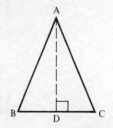

7. ABCD is a rectangle. AB = 4.2 cm, and AC = 6.3 cm.
Find CÂB and the acute angle between the diagonals.

8.   In △ABC, AC = CB = 12 cm and AB = 10 cm. Find $\widehat{A}$ and the other angles of △ABC.

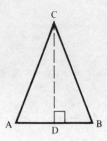

If a road gradient is 1 in 5, you rise 1 unit as you walk 5 units up the slope. The angle of slope is the angle of inclination of the road.

If the gradient of a road is given as 10%, then because $10\% = \frac{1}{10}$, the gradient is 1 in 10.

Notice that we do not find the gradient of a road in the same way as we find the gradient of a vector or a line.

The gradient of a road is 1 in 5. Find its angle of slope.

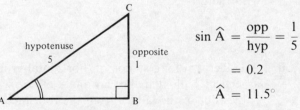

$$\sin \widehat{A} = \frac{\text{opp}}{\text{hyp}} = \frac{1}{5}$$

$$= 0.2$$

$$\widehat{A} = 11.5°$$

Therefore the angle of slope is 11.5°.

9.   What is the angle of slope when the gradient of a road is 1 in 8?

10. A road gradient is 1 in 6. What is the angle of slope?

11. Find the angle of slope of a road with a gradient of 10%. (Because $10\% = \frac{1}{10}$, a gradient of 10% is the same as a gradient of 1 in 10.)

12. Find the angle of slope of a road with a gradient of 5%.

EXERCISE 19h   **1.**   Find   a) $\sin \widehat{A}$ where $\widehat{A} = 40°$ b) $\cos \widehat{B}$ where $\widehat{B} = 50°$
What do you notice about your answers?

2.   Use the diagram to find   a) $\sin \widehat{A}$
b) $\cos \widehat{C}$
What is the value of $\widehat{A} + \widehat{C}$?

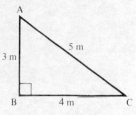

3.   If $\sin \widehat{A} = 0.3$, write down the value of $\cos \widehat{C}$.

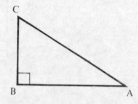

4.   In $\triangle PQR$, $\widehat{P} = 90°$, and $\cos \widehat{Q} = 0.8$. Write down the value of $\sin \widehat{R}$.

5.   Sketch a triangle ABC in which $\widehat{A} = 90°$ and $\widehat{B} = 45°$. What is the value of $\widehat{C}$? What kind of triangle is $\triangle ABC$? *Without* using a calculator or tables, write down the value of tan 45°.

MIXED EXERCISES

EXERCISE 19i   **1.**   Find the sine of 83°.

2.   If $\tan \widehat{A} = 1.6341$, find $\widehat{A}$.

3.   Find the cosine of 28°.

4.   Find $\widehat{X}$ if $\sin \widehat{X} = 0.5$

5.   Find AB.

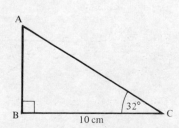

6. Find $\hat{R}$.

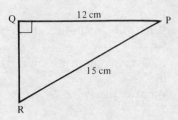

7. Find YZ.

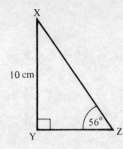

EXERCISE 19j

1. Find $\cos \hat{A}$ where $\hat{A} = 25°$

2. Find $\hat{C}$ given that $\sin \hat{C} = 0.9311$

3. Find $\tan \hat{Y}$ where $\hat{Y} = 45°$

4. Find $\hat{M}$ given that $\cos \hat{M} = 0.9311$

5. Find MN.

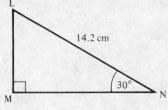

6. Find $\hat{D}$.

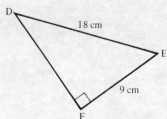

7. Find AC.

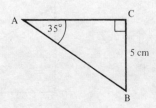

20 SQUARES AND SQUARE ROOTS

SQUARES

We obtain the *square* of a number when we multiply the number by itself.

EXERCISE 20a

> Find the square of   a) 4   b) 0.02
>
> a) $4^2 = 4 \times 4 = 16$
>
> b) $0.02^2 = 0.02 \times 0.02 = 0.0004$

Find the squares of the numbers in questions 1 to 15.

1. 3	**6.** 50	**11.** 0.3
2. 5	**7.** 300	**12.** 2000
3. 9	**8.** 0.02	**13.** 0.004
4. 30	**9.** 500	**14.** 1
5. 0.4	**10.** 10	**15.** 0.03

> Write 32 correct to 1 s.f. and use this to give a rough estimate of the square of 32.
>
> $$32 \approx 30$$
> $$32^2 \approx 30 \times 30 = 900$$

In questions 16 to 27, give each number correct to 1 s.f. then use this to give a rough estimate of the square of the number.

16. 28	**20.** 7.9	**24.** 0.0312
17. 99	**21.** 37.2	**25.** 87
18. 4.2	**22.** 1212	**26.** 0.081
19. 0.27	**23.** 73	**27.** 249

295

FINDING SQUARES

Using a calculator

Enter the number to be squared and press the "square" button, which is usually labelled x^2. If there is no "square" button, then multiply the number by itself.

Check that the answer you obtain agrees with your rough estimate. Give your answer correct to 4 significant figures unless you are told otherwise.

Using tables

To find the square of 5.28, ignore the decimal point and look up 528.

The point is then placed in the number you find from the tables in a position such that your answer agrees in size with your rough estimate. It may be necessary to fill in with zeros.

EXERCISE 20b

> Find the squares of   a) 6.29   b) 0.0341
>
>   a) $6.29^2 \approx 6 \times 6 = 36$
>
> $6.29^2 = 39.56$
>
>   b) $0.0341^2 \approx 0.03 \times 0.03 = 0.0009$
>
> $0.0341^2 = 0.001\,163$

Find the squares of:

1. 7.8	**5.** 0.16	**9.** 51.3	**13.** 1.02
2. 38	**6.** 0.032	**10.** 9.8	**14.** 13.6
3. 79.2	**7.** 48.2	**11.** 12.1	**15.** 17
4. 0.41	**8.** 11.3	**12.** 2.94	**16.** 1.11

17. 7.21	**21.** 0.879	**25.** 0.245	**29.** 0.142
18. 11.6	**22.** 0.0362	**26.** 0.072	**30.** 9.73
19. 241	**23.** 72.4	**27.** 14.2	**31.** 13.9
20. 0.824	**24.** 3.78	**28.** 142	**32.** 0.0727

33. a) Copy and complete the following table:

x	0	0.5	1	1.5	2	2.5	3	3.5	4
x^2	0			2.25	4				

b) Draw axes for x from 0 to 4 using 2 cm to 1 unit and for y from 0 to 16 using 1 cm to 1 unit. Use the table to draw the graph of $y = x^2$.

c) From the graph, find the values of y when $x = 2.2$, 1.8, 3.1 and 2.7.

d) Use a calculator or tables to find 2.2^2, 1.8^2, 3.1^2 and 2.7^2. How do these answers compare with your answers to part (c)?

e) Repeat parts (c) and (d) with other values of your own choice.

34. a) Copy and complete the following table:

x	2	4	6	8	10	12	14	15
x^2	4		36		100			225

b) Draw axes for x from 0 to 15 using 1 cm $\equiv$ 1 unit and for y from 0 to 240 using 1 cm $\equiv$ 10 units. Use your table to draw the graph of $y = x^2$.

c) From the graph, find the values of y when $x = 5.5$, 8.4, 12.8 and 13.6.

d) Use a calculator or tables to find 5.5^2, 8.4^2, 12.8^2 and 13.6^2. How do these answers compare with your answers to part (c)?

EXERCISE 20c

Find the square of 213 as accurately as possible, using a calculator or tables.

$$213^2 \approx 200 \times 200 = 40\,000$$
$$213^2 = 45\,369$$

Find the squares of the following numbers as accurately as possible, using a calculator or tables:

1.	236	**4.**	4160	**7.**	793	**10.**	68.4
2.	461	**5.**	32.4	**8.**	6240	**11.**	391
3.	5260	**6.**	321	**9.**	1430	**12.**	4690

AREAS OF SQUARES

EXERCISE 20d

Find the area of a square of side 7.2 m.

$$\text{Area} = (7.2 \times 7.2)\,\text{m}^2$$
$$\approx (7 \times 7)\,\text{m}^2 = 49\,\text{m}^2$$

$$\text{Area} = 51.8\,\text{m}^2 \text{ correct to 3 s.f.}$$

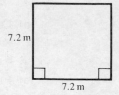

Find the areas of the squares whose sides are given in questions 1 to 9.
Give your answers correct to 3 s.f.

1.	2.4 cm	**4.**	1.06 m	**7.**	0.062 m
2.	9.6 m	**5.**	17.2 cm	**8.**	324 km
3.	32.4 cm	**6.**	52 mm	**9.**	0.31 cm

SQUARE ROOTS

The square root of a number is the number which, when multiplied
by itself, gives the original number,

e.g. since $4^2 = 16$, the square root of 16 is 4.

The square root could also be -4 since $(-4) \times (-4) = 16$ but we will
deal only with positive square roots in this chapter.

The symbol for the positive square root is $\sqrt{}$
so $\sqrt{16} = 4$

EXERCISE 20e Find the square roots in questions 1 to 18.

1.	$\sqrt{9}$	**4.**	$\sqrt{81}$	**7.**	$\sqrt{49}$
2.	$\sqrt{25}$	**5.**	$\sqrt{100}$	**8.**	$\sqrt{64}$
3.	$\sqrt{4}$	**6.**	$\sqrt{36}$	**9.**	$\sqrt{1}$

10.	$\sqrt{8100}$	**13.**	$\sqrt{4900}$	**16.**	$\sqrt{400}$
11.	$\sqrt{0.81}$	**14.**	$\sqrt{490\,000}$	**17.**	$\sqrt{2500}$
12.	$\sqrt{0.64}$	**15.**	$\sqrt{0.04}$	**18.**	$\sqrt{10\,000}$

Use the answers to Exercise 20a, questions 1 to 15, to find the following square roots.

19. $\sqrt{0.09}$ **21.** $\sqrt{0.0004}$ **23.** $\sqrt{4\,000\,000}$

20. $\sqrt{0.16}$ **22.** $\sqrt{250\,000}$ **24.** $\sqrt{0.000\,016}$

ROUGH ESTIMATES OF SQUARE ROOTS

So far, we have been able to find exact square roots of the numbers we have been given. Most numbers, however, do not have exact square roots; $\sqrt{23}$, for example, lies between 4 and 5 because $4 \times 4 = 16$, and $5 \times 5 = 25$.

$\sqrt{23}$, if given as a decimal, will start with 4.

i.e. $\sqrt{23} = 4.\,{-}{-}{-}$

EXERCISE 20f

> Find the first significant figure of the square root of 30.
> $$\sqrt{30} = 5.\,{-}{-}{-}$$
> (*Check:* $5 \times 5 = 25$)

Find the first significant figure of the square roots of the following numbers:

1. 17	**6.** 10.2	**11.** 0.20
2. 10	**7.** 85	**12.** 90
3. 38	**8.** 15	**13.** 14.2
4. 40	**9.** 4.6	**14.** 0.50
5. 3	**10.** 0.05	**15.** 5.7

Notice that $\sqrt{3} = 1.\,{-}{-}{-}$ while $\sqrt{30} = 5.\,{-}{-}{-}$

and that $\sqrt{300} = 1{-}.\,{-}{-}{-}$ while $\sqrt{3000} = 5{-}.\,{-}{-}{-}$

Every *pair of figures* added to the original number adds *one* figure to the approximate square root. We can pair off the figures from the decimal point, i.e. $\sqrt{3\vert00\vert00}$. Looking at the figure or figures in front of the first dividing line we can find the first significant figure of the square root.

Then $\sqrt{3\vert00\vert00.} = 1{-}{-}.\,{-}{-}$

                              ≈ 100 (*Check:* $100 \times 100 = 10\,000$)

and $\qquad\qquad \sqrt{30\vert 00\vert 00.} = 5--.--$

$$\approx 500 \quad (Check: \quad 500 \times 500 = 250\,000)$$

EXERCISE 20g

Find a rough value for the square root of 5280.

$$\sqrt{52\vert 80.} = 7-.--$$
$$\approx 70$$

(*Check:* $\quad 70 \times 70 = 4900$)

By finding the first significant figure of the square root, give a rough value for the square root of each of the following numbers:

1.	1400	**6.**	14 000	**11.**	396 000
2.	62 300	**7.**	3260	**12.**	396
3.	623	**8.**	41 600	**13.**	756
4.	7200	**9.**	4160	**14.**	75 600
5.	720	**10.**	14 860	**15.**	7 560 000

16.	4128	**18.**	15.26	**20.**	39.46
17.	729.4	**19.**	3.698	**21.**	394.6

FINDING SQUARE ROOTS

Using a calculator

Enter the number, say 5280, then press the square root button which is labelled $\sqrt{x}$. You will usually get a number which fills the display; give your answer correct to 4 significant figures.

$$\sqrt{5280} = 72.66$$

Check that it agrees with your rough estimate.

Using tables

Look up the number in some square root tables. There will sometimes be a choice of two numbers but the correct answer must agree with the rough estimate both in its first figure and its size.

EXERCISE 20h

Find the square root of 725 correct to 3 s.f.

$$\sqrt{7\,\vert\,25} = 2 -.-- \qquad (20 \times 20 = 400)$$

$$\sqrt{725} = 26.9 \text{ correct to 3 s.f.}$$

Find the square roots of the following numbers correct to 3 s.f. Give a rough estimate first in each case.

1.	38.4	**8.**	5.7	**15.**	10 300
2.	19.8	**9.**	650	**16.**	412 000
3.	428	**10.**	65	**17.**	728
4.	4230	**11.**	11.2	**18.**	7280
5.	32	**12.**	58	**19.**	61
6.	9.8	**13.**	24	**20.**	7 280 000
7.	67	**14.**	19	**21.**	115

22. Find the square roots of the numbers in Exercise 20g.

ROUGH ESTIMATES OF SQUARE ROOTS OF NUMBERS LESS THAN 1

$$0.2 \times 0.2 = 0.04 \qquad \text{so} \qquad \sqrt{0.04} = 0.2$$

and $\quad \sqrt{0.05} = 0.2---$ also $\sqrt{0.0004} = 0.02$

so $\quad \sqrt{0.0005} = 0.02---$ but $\sqrt{0.004}$ is neither 0.2 nor 0.02

It is easiest to find a rough estimate of the square root by again pairing off from the decimal point, but this time going to the right instead of to the left: $\sqrt{0.\vert 00 \vert 40}$, adding a zero to complete the pair.

Now $\sqrt{40} = 6.---$ so we see that $\sqrt{0.004} = 0.06---$
(*Check:* $0.06 \times 0.06 = 0.0036 \approx 0.004$)

Using a calculator or tables we find

$$\sqrt{0.004} = 0.0632 \text{ correct to 3 s.f.}$$

Note that each pair of zeros after the decimal point gives one zero after the decimal point in the answer.

EXERCISE 20i

> Find the square roots of 0.007 32 and 0.000 732 correct to 3 s.f.
>
> $$\sqrt{0.\,|\,00\,|\,73\,|\,2} = 0.08\text{---}$$
>
> $$\sqrt{0.007\,32} = 0.0856 \qquad \text{correct to 3 s.f.}$$
>
> $$\sqrt{0.\,|\,00\,|\,07\,|\,32} = 0.02\text{---}$$
>
> $$\sqrt{0.000\,732} = 0.0271 \qquad \text{correct to 3 s.f.}$$

Find a rough estimate (as far as the first significant figure) and then use your calculator or tables to find the square root of each of the following numbers correct to 3 s.f.

1.	0.042	**8.**	0.278	**15.**	0.0432
2.	0.42	**9.**	0.0278	**16.**	0.009 61
3.	0.014	**10.**	0.002 78	**17.**	0.832
4.	0.56	**11.**	0.3	**18.**	0.32
5.	0.000 14	**12.**	0.173	**19.**	0.052
6.	0.5	**13.**	0.2	**20.**	0.75
7.	0.6014	**14.**	0.69	**21.**	0.000 073

EXERCISE 20j

> Find the side of the square whose area is 50 m².
>
> Length of the side $= \sqrt{50}$ m
>
> $\qquad\qquad\qquad = 7.\text{---}$ m
>
>
>
> Length of the side = 7.07 m correct to 3 s.f.

Find the sides of the squares whose areas are given below. Give your answers correct to 3 s.f.

1.	85 cm²	**5.**	0.06 m²	**9.**	0.0085 km²
2.	120 cm²	**6.**	15.1 cm²	**10.**	59 cm²
3.	500 m²	**7.**	749 mm²	**11.**	241 m²
4.	32 m²	**8.**	84 300 km²	**12.**	61 cm²

21 PYTHAGORAS' THEOREM

We saw in a previous chapter that in a right-angled triangle there is a relationship between sides and angles. Now we can show that there is a relationship between the lengths of the three sides.

EXERCISE 21a First we will collect some evidence. Bear in mind that, however accurate your drawing, it is not perfect.

Construct the triangles in questions 1 to 6 and in each case measure the third side, the hypotenuse.

1.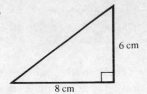

4.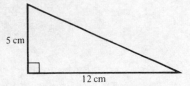

2.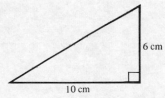

5.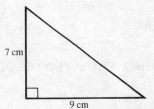

3.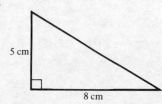

6.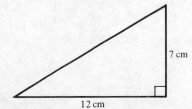

7. In each of the questions 1 to 6, find the squares of the lengths of the three sides. Write the squares in ascending size order (i.e. the smallest first). Can you see a relation between the first two squares and the third square?

303

PYTHAGORAS' THEOREM

If your drawings are reasonably accurate you will find that by adding the squares of the two shorter sides you get the square of the hypotenuse.

$$AB^2 = 16$$

$$BC^2 = 9$$

$$AC^2 = 25$$

$$25 = 16 + 9$$

so $$AC^2 = AB^2 + BC^2$$

> This result is called Pythagoras' theorem, which states that in a right-angled triangle the square of the hypotenuse is equal to the sum of the squares of the other two sides.

Pythagoras (c. 500 BC) was the Greek mathematician and philosopher who is supposed to have been the first to state the theorem in an organized way. The results, however, had already been known in Egypt and Mesopotamia for a thousand years or more.

FINDING THE HYPOTENUSE

EXERCISE 21b Give your answers correct to 3 s.f.

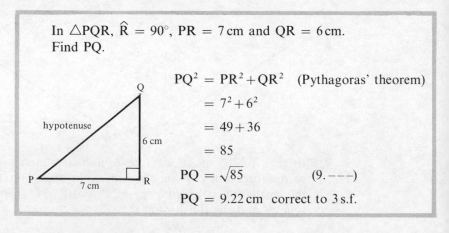

In $\triangle PQR$, $\hat{R} = 90°$, PR = 7 cm and QR = 6 cm.
Find PQ.

$$PQ^2 = PR^2 + QR^2 \quad \text{(Pythagoras' theorem)}$$

$$= 7^2 + 6^2$$

$$= 49 + 36$$

$$= 85$$

$$PQ = \sqrt{85} \qquad (9.---)$$

$$PQ = 9.22 \text{ cm} \quad \text{correct to 3 s.f.}$$

In the following right-angled triangles find the required lengths.

1. Find AC.

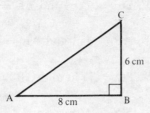

6. Find QR.

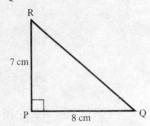

2. Find PR.

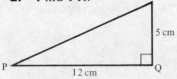

7. Find AC.

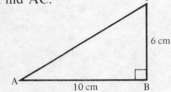

3. Find MN.

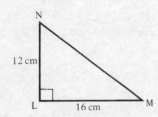

8. Find EF.

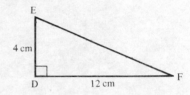

4. Find AC.

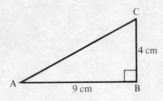

9. Find QR.

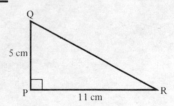

5. Find LN.

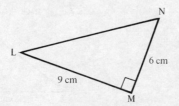

10. Find YZ.

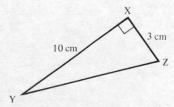

11. In $\triangle ABC$, $\hat{C} = 90°$, AC = 2 cm and BC = 3 cm. Find AB.

12. In $\triangle DEF$, $\hat{E} = 90°$, DE = 7 cm and EF = 9 cm. Find DF.

13. In $\triangle ABC$, $\hat{A} = 90°$, AB = 4 m and AC = 5 m. Find BC.

14. In $\triangle PQR$, $\hat{Q} = 90°$, PQ = 11 m and QR = 3 m. Find PR.

15. In $\triangle XYZ$, $\hat{X} = 90°$, YX = 12 m and XZ = 2 cm. Find YZ.

In $\triangle XYZ$, $\hat{Z} = 90°$, XZ = 5.3 cm and YZ = 3.6 cm. Find XY.

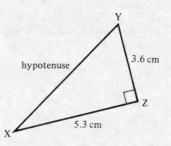

$$XY^2 = XZ^2 + ZY^2 \quad \text{(Pythagoras' theorem)}$$

$$= 5.3^2 + 3.6^2 \qquad\qquad 5.3^2 \approx 5 \times 5 = 25$$

$$= 28.09 + 12.96 \qquad\qquad 3.6^2 \approx 4 \times 4 = 16$$

$$= 41.05$$

$$XY = \sqrt{41.05} \qquad\qquad (6.\text{---})$$

$$XY = 6.41 \text{ cm} \quad \text{correct to 3 s.f.}$$

---

16. Find AC.

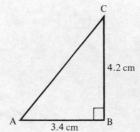

17. Find AC.

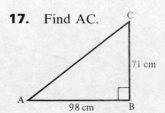

18. Find XY.

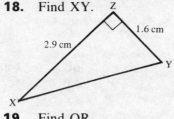

20. Find PR.

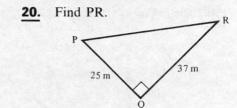

19. Find QR.

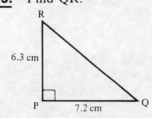

21. Find DF.

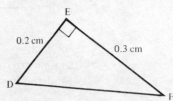

22. In △ABC, $\widehat{B} = 90°$, AB = 7.9 cm, BC = 3.5 cm. Find AC.

23. In △PQR, $\widehat{Q} = 90°$, PQ = 11.4 m, QR = 13.2 m. Find PR.

24. In △XYZ, $\widehat{Z} = 90°$, XZ = 1.23 cm, ZY = 2.3 cm. Find XY.

25. In △ABC, $\widehat{C} = 90°$, AC = 32 cm, BC = 14.2 cm. Find AB.

26. In △PQR, $\widehat{P} = 90°$, PQ = 9.6 m, PR = 8.8 m. Find QR.

27. In △DEF, $\widehat{F} = 90°$, DF = 10.1 cm, EF = 6.4 cm. Find DE.

THE 3,4,5 TRIANGLE

You will have noticed that, in most cases when two sides of a right-angled triangle are given and the third side is calculated using Pythagoras' theorem, the answer is not an exact number. There are a few special cases where all three sides are exact numbers.

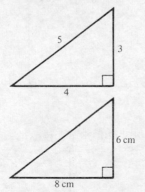

The simplest one is the 3,4,5 triangle. Any triangle similar to this has sides in the ratio 3 : 4 : 5 so whenever you spot this case you can find the missing side very easily.

For instance, in the triangle opposite, $6 = 2 \times 3$ and $8 = 2 \times 4$. The triangle is similar to the 3,4,5 triangle, so the hypotenuse is 2×5 cm, that is, 10 cm.

The other triangle with exact sides which might be useful is the 5,12,13 triangle.

EXERCISE 21c

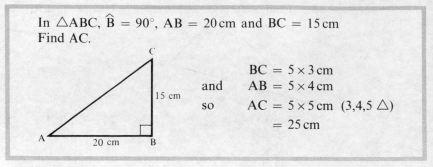

In $\triangle ABC$, $\widehat{B} = 90°$, $AB = 20$ cm and $BC = 15$ cm
Find AC.

    BC $= 5 \times 3$ cm
and AB $= 5 \times 4$ cm
so   AC $= 5 \times 5$ cm   (3,4,5 $\triangle$)
        $= 25$ cm

In each of the following questions, decide whether the triangle is similar to the 3,4,5 triangle or to the 5,12,13 triangle or to neither. Find the hypotenuse, using the method you think is easiest.

1.

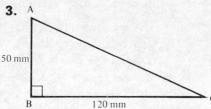

5.

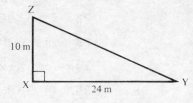

2.

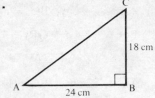

6.

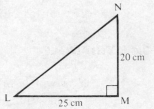

3.

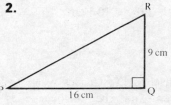

7.

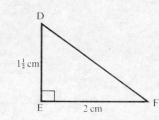

4.

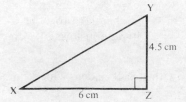

8.

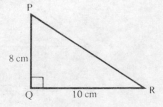

FINDING ONE OF THE SHORTER SIDES

If we are given the hypotenuse and one other side we can find the third side.

EXERCISE 21d

In $\triangle ABC$, $\hat{B} = 90°$, $AB = 7\,cm$ and $AC = 10\,cm$.
Find BC.

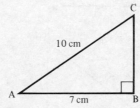

$$AC^2 = BC^2 + AB^2 \quad \text{(Pythagoras' theorem)}$$
$$10^2 = BC^2 + 7^2$$
$$100 = BC^2 + 49$$
$$51 = BC^2 \quad \text{(taking 49 from both sides)}$$
$$BC = \sqrt{51} \quad (7.\text{---})$$
$$BC = 7.14\,cm \quad \text{correct to 3 s.f.}$$

1. Find BC.

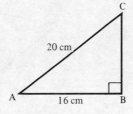

3. Find PQ.

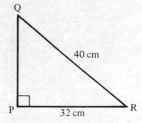

2. Find LM.

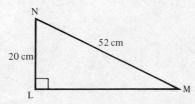

4. Find YZ.

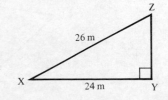

Give your answers to questions 5 to 14 correct to 3 s.f.

5. Find BC.

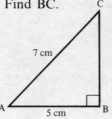

6. Find RQ.

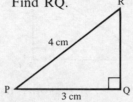

7. Find AB.

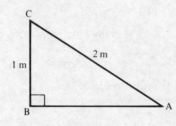

8. Find AB.

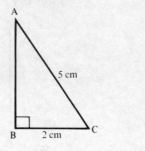

9. Find EF.

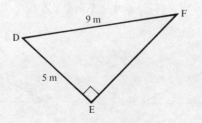

10. Find BC.

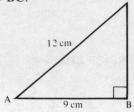

11. Find XY.

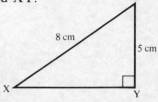

12. Find QR.

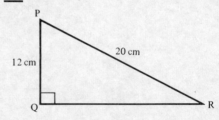

13. Find XY.

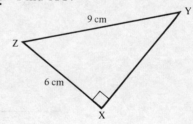

14. Find PQ.

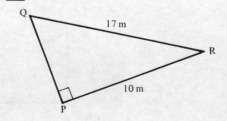

MIXED EXAMPLES

EXERCISE 21e In each case find the length of the missing side. If any answers are not exact give them correct to 3 s.f.

If you notice a 3,4,5 triangle or a 5,12,13 triangle, you can use it to get the answer quickly.

1. Find AC.

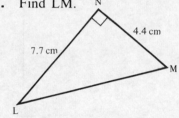

2. Find LM.

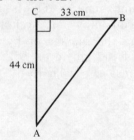

3. Find AB.

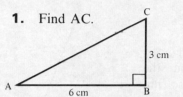

4. Find PR.

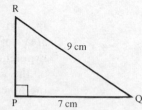

5. Find DF.

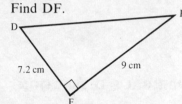

6. Find YZ.

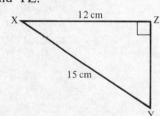

7. In $\triangle$ABC, $\hat{B} = 90°$, AB = 2 cm, AC = 4 cm. Find BC.

8. In $\triangle$ABC, $\hat{B} = 90°$, AB = 1.25 m, CA = 8.25 m. Find BC.

9. In $\triangle$PQR, $\hat{Q} = 90°$, PQ = 65 cm, QR = 60 cm. Find PR.

10. In $\triangle$DEF, $\hat{D} = 90°$, DE = 124 cm, DF = 234 cm. Find EF.

11. In $\triangle$ABC, $\hat{C} = 90°$, AC = 3.2 cm, AB = 9.81 cm. Find BC.

12. In $\triangle$XYZ, $\hat{Y} = 90°$, XY = 1.5 cm, YZ = 2 cm. Find XZ.

13. In $\triangle$PQR, $\hat{P} = 90°$, PQ = 5.1 m, QR = 8.5 m. Find PR.

14. In $\triangle$ABC, $\hat{C} = 90°$, AB = 92 cm, BC = 21 cm. Find AC.

15. In $\triangle$XYZ, $\hat{X} = 90°$, XY = 3.21 m, XZ = 1.43 m. Find YZ.

PYTHAGORAS' THEOREM USING AREAS

The area of a square is found by squaring the length of its side, so we can represent the squares of numbers by areas of squares.

This gives us a version of Pythagoras' theorem, using areas:

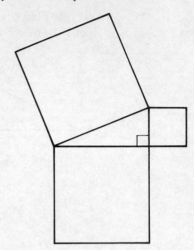

In a right-angled triangle, the area of the square on the hypotenuse is equal to the sum of the areas of the squares on the other two sides.

PERIGAL'S DISSECTION

On squared paper, and using 1 cm to 1 unit, copy the left-hand diagram. Make sure that you draw an accurate square on the hypotenuse either by counting the squares or by using a protractor and a ruler. D is the centre of the square on AB. Draw a vector $\overrightarrow{DE}$ so that $\overrightarrow{DE} = \frac{1}{2}\overrightarrow{AC}$, i.e. DE must be parallel to AC.

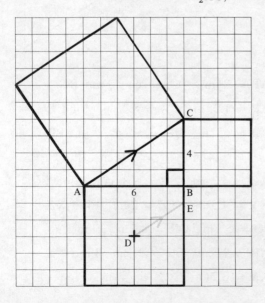
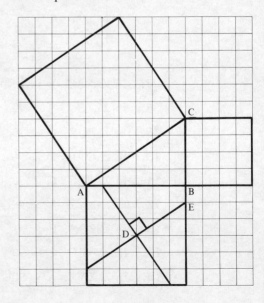

Now complete the drawing as in the right-hand diagram. Make sure that the angles at D are right angles.

Cut out the smallest square and the four pieces from the middle-sized square. These five pieces can be fitted exactly, like a jigsaw, into the outline of the biggest square.

FINDING LENGTHS IN AN ISOSCELES TRIANGLE

An isosceles triangle can be split into two right-angled triangles and this can sometimes help when finding missing lengths, as it did when finding angles.

EXERCISE 21f

In $\triangle ABC$, $AB = BC = 12$ cm and $AC = 8$ cm.
Find the height of the triangle.

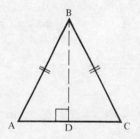 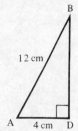

(Join B to D, the mid point of AC)

$$AB^2 = AD^2 + BD^2 \quad \text{(Pythagoras' theorem)}$$
$$12^2 = 4^2 + BD^2$$
$$144 = 16 + BD^2$$
$$128 = BD^2 \quad \text{(taking 16 from both sides)}$$
$$BD = \sqrt{1\,28.} \qquad (1 - . - -)$$
$$BD = 11.3 \text{ cm}$$

The height of the triangle is 11.3 cm correct to 3 s.f.

Give your answers correct to 3 s.f.

1.

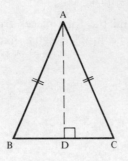

AB = AC = 16 cm. BC = 20 cm.
Find the height of the triangle.

2.

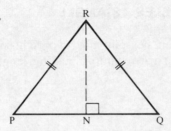

PQ = 12 cm. PR = RQ.
The height of the triangle is 8 cm.
Find PR.

3.

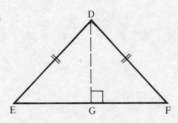

DE = DF = 20 cm. The height of
the triangle is 13.2 cm.
Find EG and hence EF.

4. In △ABC, AB = BC = 5.2 cm and AC = 6 cm. Find the
height of the triangle.

5. In △PQR, PQ = QR = 9 cm and the height of the triangle is
7 cm. Find the length of PR.

FINDING THE DISTANCE OF A CHORD FROM THE CENTRE OF A CIRCLE

AB is a chord of a circle with centre O.
OA and OB are radii and so are equal.
Hence triangle OAB is isosceles and we
can divide it through the middle into two
right-angled triangles.

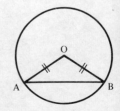

EXERCISE 21g

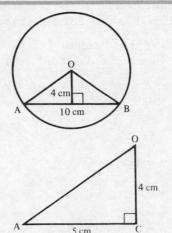

A chord AB of a circle with centre O is 10 cm long. The chord is 4 cm from O. Find the radius of the circle.

(The distance from the centre is the perpendicular distance so OC = 4 cm. From symmetry AC = 5 cm.)

$$OA^2 = AC^2 + OC^2 \quad \text{(Pythagoras' theorem)}$$
$$= 5^2 + 4^2$$
$$= 25 + 16$$
$$= 41$$
$$OA = \sqrt{41} \qquad (6.\text{---})$$
$$OA = 6.40 \text{ cm}$$

The radius of the circle is 6.40 cm correct to 3 s.f.

Give your answers correct to 3 s.f.

1. A circle with centre O has a radius of 5 cm. AB = 8.4 cm. Find the distance of the chord from the centre of the circle.

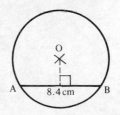

2. O is the centre of the circle and AB is a chord of length 7.2 cm. The distance of the chord from O is 3 cm. Find the radius of the circle.

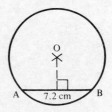

3. In a circle with centre O, a chord AB is of length 7 cm. The radius of the circle is 11 cm. Find the distance of the chord from O.

4. In a circle with centre O and radius 17 cm, a chord AB is of length 10.4 cm. Find the distance of the chord from O.

5. In a circle with centre P and radius 7.6 cm, a chord QR is 4.2 cm from P. Find the length of the chord.

PROBLEMS USING PYTHAGORAS' THEOREM

EXERCISE 21h

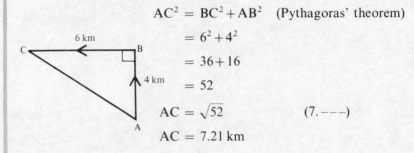

A man starts from A and walks 4 km due north to B, then 6 km due west to C. Find how far C is from A.

$$AC^2 = BC^2 + AB^2 \quad \text{(Pythagoras' theorem)}$$
$$= 6^2 + 4^2$$
$$= 36 + 16$$
$$= 52$$
$$AC = \sqrt{52} \qquad (7.\text{---})$$
$$AC = 7.21 \text{ km}$$

The distance of C from A is 7.21 km, correct to 3 s.f.

Give your answers correct to 3 s.f.

1. A ladder 3 m long is leaning against a wall. Its foot is 1.5 m from the foot of the wall. How far up the wall does the ladder reach?

2. ABCD is a rhombus. AC = 10 cm and BD = 12 cm. Find the length of a side of the rhombus.

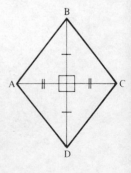

3. Find the length of a diagonal of a square of side 10 cm.

4. A hockey pitch measures 55 m by 90 m. Find the length of a diagonal of the pitch.

5. A wire stay 11 m long is attached to a telegraph pole at a point A, 8 m up from the ground. The other end of the stay is fixed to a point B, on the ground. How far is B from the foot of the telegraph pole?

6.

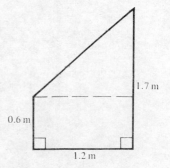

In the kite ABCD, $\hat{A} = \hat{C} = 90°$. DC = 41 cm and BC = 62 cm. Find the length of the diagonal BD.

7. A diagonal of a football pitch is 130 m long and the long side measures 100 m. Find the length of the short side of the pitch.

8. The diagram shows the side view of a coal bunker. Find the length of the slant edge.

1.7 m
0.6 m
1.2 m

9.

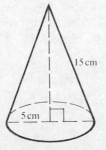

15 cm
5 cm

The slant height of a cone is 15 cm and the base radius is 5 cm. Find the height of the cone.

10. A man starts from A and walks 6.5 km due south to B; then he walks due east to C. He is then 9 km from A.
How far is C from B?

PROBLEMS USING PYTHAGORAS' THEOREM AND TRIGONOMETRY

EXERCISE 21i

A rectangle measures 13 cm by 7 cm. Find the length of a diagonal and the angle between this diagonal and the shorter side.

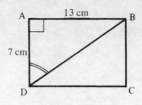

$$BD^2 = AD^2 + AB^2 \quad \text{(Pythagoras' theorem)}$$
$$= 13^2 + 7^2$$
$$= 169 + 49$$
$$= 218$$
$$BD = \sqrt{2\vdots 18}. \qquad (1-.--)$$
$$BD = 14.8\,\text{cm} \text{ correct to 3 s.f.}$$

$$\tan A\hat{D}B = \frac{\text{opp}}{\text{adj}} = \frac{13}{7}$$
$$= 1.857$$
$$A\hat{D}B = 61.7^\circ$$

The diagonal is 14.8 cm long correct to 3 s.f. and the angle between the diagonal and the shorter side is 61.7°.

Give your answers correct to 3 s.f.

1. In $\triangle ABC$, $AB = BC = 10$ cm and $AC = 14$ cm. Find the height of the triangle and its angles.

2. A ladder 4 m long leans against a wall so that its top is 3 m up the wall. Find how far out from the wall the foot of the ladder is and find the angle the ladder makes with the wall.

3. A man starts from A and drives 16 km due west to B, then due south 10 km to C. How far is C from A and what is the bearing of A from C?

4. The diagonals of a rhombus are 18 cm and 27.4 cm long. Find the sides and angles of the rhombus.

5. R is a point 20 m from the foot, Q, of a pole. The angle of elevation of the top of the pole from R is 28°. Find the height of the pole and the distance of R from P.

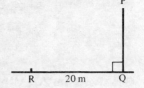

6. Town A is 22 km due west of town B. Town C is 16 km due south of B.

a) Find how far A is from C.

b) Find AĈB and hence find the bearings of A from C and of C from A.

7. In rectangle ABCD, AB = 16 cm and BC = 20 cm. E is a point on BC such that BE = 8 cm. Find how far E is from A and from D.

8. Sketch axes for *x* and *y* from 0 to 8. A is the point (1, 2) and B is (6, 8). Find the length of AB.

9. Sketch axes for *x* and *y* from −4 to 4. P is the point (−2, 4) and Q is (3, −1). Find the length PQ.

10. Sketch axes for *x* and *y* from −6 to 6. R is the point (−6, −6) and S is the point (6, −2). Find the length RS.

22 PRACTICAL APPLICATIONS OF GRAPHS

GRAPHS INVOLVING STRAIGHT LINES

If you were to go to Spain for a holiday, you would probably have a little difficulty in knowing the cost of things in pounds and pence. If we know the rate of exchange, we can use a simple straight line graph to convert a given number of pesetas into pounds or a given number of pounds into pesetas.

Given that £1 converts to 210 pesetas, we can draw a graph to convert values from, say, £0–£90 into pesetas. Take 2 cm ≡ £10 and 1 cm ≡ 1000 pesetas. (≡ means "is equivalent to".)

Since   £1 ≡ 210 pesetas

                         £10 ≡ 2100 pesetas

and   £60 ≡ 12 600 pesetas

We now plot these points and join them with a straight line.

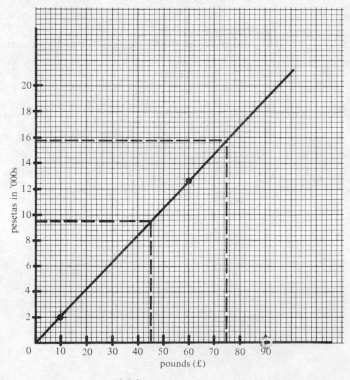

From the graph:

$$£45 \equiv 9450 \text{ pesetas}$$
$$£72 \equiv 15\,100 \text{ pesetas}$$
$$6400 \text{ pesetas} \equiv £30.50$$
$$15\,800 \text{ pesetas} \equiv £75$$

EXERCISE 22a **1.** The table gives temperatures in degrees Fahrenheit (°F) and the equivalent values in degrees Centigrade (°C).

Temperature in °F	57	126	158	194
Temperature in °C	14	52	70	90

Plot these points on a graph for Centigrade values from 0 to 100 and Fahrenheit values from 0 to 220. Let 2 cm represent 20 units on each axis.

Use your graph to convert:
a) 97°F into °C
b) 172°F into °C
c) 25°C into °F
d) 80°C into °F

2. The table shows the conversion from US dollars to £s for various amounts of money.

US dollars	50	100	200
£s	35	70	140

Plot these points on a graph and draw a straight line to pass through them. Let 4 cm represent 50 units on both axes.

Use your graph to convert:
a) 160 dollars into £s
b) 96 dollars into £s
c) £122 into dollars
d) £76 into dollars

3. The table shows the conversion of various sums of money from Deutschmarks to French francs.

Deutschmarks (DM)	100	270	350
French francs (f)	310	837	1085

Plot these points on a graph and draw a straight line to pass through them. Take 2 cm to represent 50 units on the DM-axis and 100 units on the f-axis.

Use your graph to convert:

a) 160 DM into francs

b) 330 DM into francs

c) 440 f into Deutschmarks

d) 980 f into Deutschmarks

4. The table shows the distance a girl walks in a given time.

Time walking in hours	0	1	$2\frac{1}{2}$	4	5
Distance walked in km	0	6	15	24	30

Draw a graph of these results. What do you conclude about the speed at which she walks?

How far has she walked in   a) 2 hours   b) $3\frac{1}{2}$ hours?

How long does she take to walk   c) 10 km   d) 21 km?

5. The table shows the distance an aircraft has travelled at various times on a particular journey.

Time after departure in hours	0	1	$3\frac{1}{2}$	6
Distance travelled from take-off in km	0	550	1925	3300

Draw a graph of these results. What can you conclude about the speed of the aircraft?

How far does it fly in   a) $1\frac{1}{2}$ hours   b) $4\frac{1}{2}$ hours?

How long does it take to fly   c) 1000 km   d) 2500 km?

6. Marks in an examination range from 0 to 65. Draw a graph which enables you to express the marks in percentages from 0 to 100. Note that a mark of 0 is 0% while a mark of 65 is 100%.

Use your graph   a) to express marks of 35 and 50 as percentages

b) to find the original mark for percentages of 50% and 80%.

7. Deductions from the wages of a group of employees amount to £35 for every £100 earned. Draw a graph to show the deductions made from gross pay in the range £0–£400 per week.

How much is deducted from an employee whose gross weekly pay is a) £125 b) £240 c) £335? How much is earned each week by an employee whose weekly deductions amount to d) £40 e) £88?

8. The table shows the fuel consumption figures for a car in both miles per gallon *(X)* and in kilometres per litre *(Y)*.

mpg *(X)*	30	45	60
km/litre *(Y)*	10.5	15.75	21

Plot these points on a graph taking $2\,cm \equiv 10\,units$ on the *X*-axis and $4\,cm \equiv 5\,units$ on the *Y*-axis. Your scale should cover 0–70 for *X* and 0–25 for *Y*.

Use your graph to find:

a) 12 km/litre in mpg   c) 22.5 km/litre in mpg

b) 64 mpg in km/litre   d) 23 mpg in km/litre

9. The table gives various speeds in kilometres per hour with the equivalent values in metres per second.

Speed in km/h *(S)*	0	80	120	200
Speed in m/s *(V)*	0	22.2	33.3	55.5

Plot these values on a graph taking $4\,cm \equiv 50\,units$ on the *S*-axis and $4\,cm \equiv 10\,units$ on the *V*-axis.

Use your graph to convert:

a) 140 km/h into m/s   c) 18 m/s into km/h

b) 46 m/s into km/h   d) 175 km/h into m/s

10. A number of rectangles, measuring *l* cm by *b* cm, all have a perimeter of 24 cm. Copy and complete the following table:

l	1	2	3	4	6	8
b			9			4

Draw a graph of these results using your own scale. Use your graph to find *l* if *b* is a) 2.5 cm b) 6.2 cm and to find *b* if *l* is c) 5.5 cm d) 2.8 cm

GRAPHS INVOLVING CURVES

When two quantities that are related are plotted one against the other, we often find that the points do not lie on a straight line. They may, however, lie on a smooth curve.

Consider the table below which gives John's height on his birthday over a period of 8 years.

Age in years	11	12	13	14	15	16	17	18	19
Height in cm	138	140	144	150	158	165	170	172	173

These points can be plotted on a graph and joined to give a smooth curve through the points as shown.

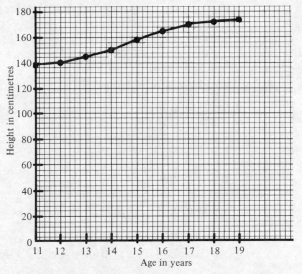

The graph enables us to estimate that:

a) he was 162 cm tall when he was $15\frac{1}{2}$ years old,

b) he was 146 cm tall when he was 13 years 5 months,

c) when he was 17 years 6 months, he was 171 cm tall.

We can also deduce that:

i) the fastest increase in height was between his fourteenth and fifteenth birthdays—the curve is steepest between these two birthdays,

ii) he grew very little between his eighteenth and nineteenth birthdays—the curve is quite flat in this region.

We could obtain more accurate results if we took 100 cm as the lowest height on the vertical axis and used a larger scale.

EXERCISE 22b   **1.**   The weights of lead spheres of various diameters are shown in the table.

Diameter in mm *(D)*	4	5.2	6.4	7.2	7.9	8.8
Weight in grams *(W)*	380	840	1560	2230	2940	4070

Plot this information on a graph and draw a smooth curve through the points. Use 2 cm ≡ 1 unit on the *D*-axis and 2 cm ≡ 500 units on the *W*-axis.

Use your graph to estimate

a) the weight of a lead sphere of diameter 6 mm

b) the diameter of a lead sphere of weight 2 kg

2.   Recorded speeds of a motorcar at various times after starting from rest are shown in the table.

Time in seconds	0	5	10	15	20	25	30	35	40
Speed in km/h	0	62	112	148	172	187	196	199	200

Taking 2 cm ≡ 5 sec and 1 cm ≡ 10 km/h, plot these results and draw a smooth curve to pass through these points.

Use your graph to estimate

a) the time which passes before the car reaches
   i) 100 km/h   ii) 150 km/h

b) its speed after   i) 13 sec   ii) 27 sec

3.   The weight of a puppy at different ages is given in the table.

Age in days *(A)*	10	20	40	60	80	100	120	140
Weight in grams *(W)*	50	100	225	425	750	875	950	988

Draw a graph to represent this data, taking 1 cm ≡ 10 days on the *A*-axis and 1 cm ≡ 50 g on the *W*-axis.

Hence estimate

a) the weight of the puppy after   i) 50 days   ii) 114 days

b) the age of the puppy when it weighs   i) 500 g   ii) 1000 g

c) the weight it puts on between day 25 and day 55

d) its birth weight

4. The speed of a particle (v metres per second) at various times (t seconds) after starting is given in the table.

t	0	1	2	3	4	5	6	7
v	0	35	60	76.5	83	83	76	57

Plot this information on a graph using $2\,cm \equiv 1\,unit$ on the t-axis and $2\,cm \equiv 10\,units$ on the v-axis.

Use your graph to find:

a) the greatest speed of the particle and the time at which it occurs

b) its speed after   i) 3.5 sec   ii) 6.8 sec

c) when its speed is 65 m/sec

5. The cost of fuel ($£C$) per nautical mile for a ship travelling at various speeds (v knots) is given in the table.

v	12	14	16	18	20	22	24	26	28
C	18.15	17.16	16.67	16.5	16.5	16.67	16.94	17.36	17.82

Draw a graph to show how cost changes with speed. Use $1\,cm \equiv 1\,knot$ and $10\,cm \equiv £1$. (Take £16 as the lowest value for C.)

Use your graph to estimate:

a) the most economical speed for the ship and the corresponding cost per nautical mile

b) the speeds when the cost per nautical mile is £17

c) the cost when the speed is   i) 13 knots   ii) 24.4 knots

6. Cubes made from a certain metal with edges of the given lengths have weights as given in the table.

Length of edge in cm (L)	1	2	3	4	5	6
Weight of cube in grams (W)	9	72	243	576	1125	1944

Plot this information on a graph, joining the points with a smooth curve. Take $2\,cm \equiv 1\,unit$ on the L-axis and $1\,cm \equiv 100\,g$ on the W-axis.

From your graph find:

a) the weight in grams of a cube with edge
   i) 3.5 cm   ii) 5.3 cm

b) the length of the edge of a cube with weight
   i) 500 g   ii) 1500 g

7. The temperatures, taken at 2-hourly intervals, at my home on a certain day were as given in the table.

Time	Temperature in °C
midnight	4.4
2 a.m.	3.4
4 a.m.	3
6 a.m.	3.4
8 a.m.	4.8
10 a.m.	7.6
noon	11
2 p.m.	13.4
4 p.m.	14.2
6 p.m.	13.6
8 p.m.	12
10 p.m.	9
midnight	5.4

Draw a graph to show this data taking 1 cm ≡ 1 hour and 1 cm ≡ 1°C.

Use your graph to estimate:

a) the temperature at 11 a.m. and at 11 p.m.

b) the times at which the temperature was 10°C

8 The time of sunset at Greenwich on different dates, two weeks apart, is given in the table.

	Oct	Nov		Dec		Jan			Feb
Date *(D)*	23	6	20	4	18	1	15	29	12
Time *(T)*	1752	1626	1605	1554	1552	1603	1622	1646	1711

Using 1 cm ≡ 1 week on the *D*-axis and 4 cm ≡ 1 hour on the *T*-axis, plot these points on a graph and join them with a smooth curve. Take 1400 as the lowest value for *T*.

From your graph estimate

a) the time of sunset on 9 January

b) the date(s) in November when the sun sets at 1615

9. The time of sunset at Greenwich on different dates, each two weeks apart, is given in the table.

	May		June		July		Aug	
Date *(D)*	15	29	12	26	10	24	7	21
Time *(T)*	2045	2105	2118	2122	2116	2101	2039	2012

Using $1\,cm \equiv 1\,week$ on the *D*-axis and $8\,cm \equiv 1\,hour$ on the *T*-axis, plot these points on a graph and join them with a smooth curve. Take 1900 as the lowest value of *T*.

From your graph estimate:

a) the time of sunset on 17 July

b) the date on which the sun sets at 2027

10. A rectangle measuring *l* cm by *b* cm has an area of $24\,cm^2$. The table gives different values of *l* with the corresponding values of *b*.

l	1	2	3	4	6	8	12	16
b	24		8		4		2	1.5

Complete the table and draw a graph to show this information, joining the points with a smooth curve. Take $1\,cm \equiv 1\,unit$ on the *l*-axis and $1\,cm \equiv 2\,units$ on the *b*-axis.

Use your graph to estimate the value of

a) *l* when *b* is   i) 14 cm   ii) 2.4 cm

b) *b* when *l* is   i) 18 cm   ii) 2.8 cm

23 AVERAGES

We are frequently looking for ways of representing a set of figures in a simple form. Can we choose a single number that will adequately represent a set of numbers?

We try to do this by using averages.

Three different types of averages are used, each with its own individual advantages and disadvantages.

They are the *arithmetic average* or *mean*, the *mode* and the *median*.

THE ARITHMETIC AVERAGE OR MEAN

Consider a group of five children. When they are asked to produce the money they are carrying the amounts collected are 56 p, £1.42, 96 p, 24 p and 77 p respectively. If the total value of this money (£3.95) is shared equally amongst the five children, each will receive 79 p. This is called the arithmetic average or mean of the five amounts.

> The arithmetic average or mean of a set of figures is the sum of the figures divided by the number of figures in the set.

For example, the average or mean of 12, 15, 25, 42 and 16 is

$$\frac{12+15+25+42+16}{5} = \frac{110}{5} = 22$$

One commonplace use of the arithmetic average is to compare the marks of pupils in a group or form. The pupils are given positions according to their average mark over the full range of subjects they study. An advantage is that we can compare the results of pupils who study 7 subjects with those who study 11 subjects. A disadvantage is that one very poor mark may pull the mean down significantly.

The mean may also be rather artificial, for example, giving $5\frac{1}{3}$ p to each of a group of people, or having a mean shoe size of 5.1, or a mean family size of 2.24 children.

329

EXERCISE 23a Find the arithmetic average or mean of the following sets of numbers:

1. 3, 6, 9, 14
2. 2, 4, 9, 13
3. 12, 13, 14, 15, 16, 17, 18
4. 23, 25, 27, 29, 31, 33, 35
5. 19, 6, 13, 10, 32
6. 34, 14, 39, 20, 16, 45

7. 1.2, 2.4, 3.6, 4.8
8. 18.2, 20.7, 32.5, 50, 78.6
9. 6.3, 4.5, 6.8, 5.2, 7.3, 7.1
10. 3.1, 0.4, 7.2, 0.7, 6.1
11. 38.2, 17.6, 63.5, 80.7
12. 0.76, 0.09, 0.35, 0.54, 1.36

John's examination percentages in 8 subjects were 83, 47, 62, 49, 55, 72, 58 and 62. What was his mean mark?

Mean mark for 8 subjects

$$= \frac{83+47+62+49+55+72+58+62}{8}$$

$$= \frac{488}{8}$$

$$= 61$$

---

13. In the Christmas terminal examinations Lisa scored a total of 504 in 8 subjects. Find her mean mark.

14. A darts player scored 2304 in 24 visits to the board. What was his average number of points per visit?

15. A bowler took 110 wickets for 1815 runs. Calculate his average number of runs per wicket.

16. Peter's examination percentages in 7 subjects were 64, 43, 86, 74, 55, 53 and 66. What was his mean mark?

17. In six consecutive English examinations, Jane's percentage marks were 83, 76, 85, 73, 64 and 63. Find her mean mark.

18. A football team scored 54 goals in 40 league games. Find the average number of goals per game.

19. The first Hockey XI scored 14 goals in their first 16 matches. What was the average number of goals per match?

20. In an ice-dancing competition the recorded scores for the winners were 5.8, 5.9, 6.0, 5.8, 5.8, 5.8, 5.6 and 5.7. Find their mean score.

21. The recorded rainfall each day at a holiday resort during the first week of my holiday was 3 mm, 0, 4.5 mm, 0, 0, 5 mm and 1.5 mm. Find the mean daily rainfall for the week.

22. The weights of the members of a rowing eight were 82 kg, 85 kg, 86 kg, 86 kg, 84 kg, 88 kg, 92 kg and 85 kg. Find the average weight of the "eight". If the cox weighed 41 kg, what was the average weight of the crew?

On average my car travels 28.5 miles on each gallon of petrol. How far will it travel on 30 gallons?

If the car travels 28.5 miles on 1 gallon of petrol it will travel 30×28.5 miles, i.e. 855 miles, on 30 gallons.

---

23. My father's car travels on average 33.4 miles on each gallon of petrol. How far will it travel on 55 gallons?

24. Olga's car travels on average 12.6 km on each litre of petrol. How far will it travel on 205 litres?

25. The average daily rainfall in Puddletown during April was 2.4 mm. How much rain fell during the month?

26. The daily average number of hours of sunshine during my 14 day holiday in Greece was 9.4. For how many hours did the sun shine while I was on holiday?

Elaine's average mark after 7 subjects is 56 and after 8 subjects it has risen to 58. How many does she score in her eighth subject?

Total scored in 7 subjects is $56 \times 7 = 392$

Total scored in 8 subjects is $58 \times 8 = 464$

Score in her eighth subject
$$= \text{total for 8 subjects} - \text{total for 7 subjects}$$
$$= 464 - 392$$
$$= 72$$

Therefore Elaine scores 72 in her eighth subject.

27. David Gower's batting average after 11 completed innings was 62. After 12 completed innings it had increased to 68. How many runs did he score in his twelfth innings?

28. Richard was collecting money for a charity. The average amount collected from the first 15 houses at which he called was 30 p, while the average amount collected after 16 houses was 35 p. How much did he collect from the sixteenth house?

29. After six examination results Tom's average mark was 57. His next result increased his average to 62. What was his seventh mark?

30. Anne's average mark after 8 results was 54. This dropped to 49 when she received her ninth result which was for French. What was her French mark?

In five consecutive frames in the World Championships, a snooker player scored 62, 0, 13, 92 and 53. Find his average score per frame. How many did he score in the next frame if his average increased to 57?

$$\text{Average score for 5 frames} = \frac{62 + 0 + 13 + 92 + 53}{5}$$

$$= \frac{220}{5}$$

$$= 44$$

If the average score after 6 frames is 57:

total scored in 6 frames $= 57 \times 6 = 342$

But the total scored in 5 frames $= 220$

$\therefore$ score in sixth frame

$= $ total score for 6 frames $-$ total score for 5 frames

$= 342 - 220$

$= 122$

Therefore the sixth frame score was 122.

31. In seven consecutive innings a batsman scored 53, 4, 73, 104, 66, 44 and 83. What was his average? What does he score in his next innings if his average falls to 56?

32. During a certain week the number of lunches served in a school canteen were: Monday 213, Tuesday 243, Wednesday 237 and Thursday 239. Find the average number of meals served daily over the four days. If the daily average for the week (Monday–Friday) was 225, how many meals were served on Friday?

33. A paperboy's sales during a certain week were: Monday 84, Tuesday 112, Wednesday 108, Thursday 95 and Friday 131. Find his average daily sales. When he included his sales on Saturday his daily average increased to 128. How many papers did he sell on Saturday?

34. The number of hours of sunshine in Rhodes for successive days during a certain week were 10.9, 11.9, 9.9, 7.7, 11.7, 9.3 and 12.1. Find the daily average.
The following week the daily average was 11 hours. How many more hours of sunshine were there the second week than the first?

35. Jean's marks in the end of term examinations were 46, 80, 59, 83, 54, 67, 79, 82 and 62. Find her average mark. It was found that there had been an error in her mathematics mark. It should have been 74, not 83. What difference did this make to her average?

36. The heights of the 11 girls in a hockey team are 162 cm, 152 cm, 166 cm, 149 cm, 153 cm, 165 cm, 169 cm, 145 cm, 155 cm, 159 cm and 163 cm. Find the average height of the team. If the girl who was 145 cm tall were replaced by a girl 156 cm tall, what difference would this make to the average height of the team?

37. During the last five years the distances I travelled in my car, in miles, were 10 426, 12 634, 11 926, 14 651 and 13 973. How many miles did I travel in the whole period? What was my yearly average? How many miles should I travel this year to reduce the average annual mileage over the six years to 11 984?

38. The average weight of the 18 boys in a class is 63.2 kg. When two new boys join the class the average weight increases to 63.7 kg. What is the combined weight of the two new boys?

In a rugby XV the average weight of the eight forwards is 85 kg and the average weight of the seven backs is 70 kg. Find the average weight of the team.

$$\text{Total weight of 8 forwards} = 85 \times 8 \text{ kg} = 680 \text{ kg}$$

$$\text{Total weight of 7 backs} = 70 \times 7 \text{ kg} = 490 \text{ kg}$$

$\therefore$ total weight of the 15 members of the team

$$= (680 + 490) \text{ kg}$$

$$= 1170 \text{ kg}$$

$\therefore$ average weight of the team $= \dfrac{1170}{15} \text{ kg}$

$$= 78 \text{ kg}$$

39. The average height of the 12 boys in a class is 163 cm and the average height of the 18 girls is 159 cm. Find the average height of the class.

40. The average weight of the 15 girls in a class is 54.4 kg while the average weight of the 10 boys is 57.4 kg. Find the average weight of the class.

41. In a school the average size of the 14 lower school forms is 30, the average size of the 16 middle school forms is 25 and the average size of the 20 upper school forms is 24. Find the average size of form for the whole school.

42. Northshire has an area of 400 000 hectares and last year the annual rainfall was 274 cm, while Southshire has an area of 150 000 hectares and last year the annual rainfall was 314 cm. What was the annual rainfall last year for the combined area of the two counties?

43. After playing 10 three-day matches and 8 one-day matches, the average *daily* attendances for a County Cricket club were 2160 for three-day matches and 4497 for one-day matches. Calculate the average *daily* attendance for the 18 matches.

MODE

The mode of a set of numbers is the number that occurs most frequently, e.g. the mode of the numbers 6, 4, 6, 8, 10, 6, 3, 8 and 4 is 6, since 6 is the only number occurring more than twice.

It would obviously be of use for a firm with a chain of shoe shops to know that the mode or modal size for men's shoes in one part of the country is 8, whereas in another part of the country it is 7. Such information would influence the number of pairs of shoes of each size kept in stock.

If all the figures in a set of figures are different, there cannot be a mode, for no figure occurs more frequently than all the others. On the other hand, if two figures are equally the most popular, there will be two modes.

In Book 1, Chapter 22, we used bar charts to show such things as the spread of heights in a group of children, and the favourite colour of a group of people. These may be used to determine the mode of the group.

The following bar chart is reproduced from Book 1. It shows the colour selected by 35 people when asked to choose their favourite colour from a card showing six colours.

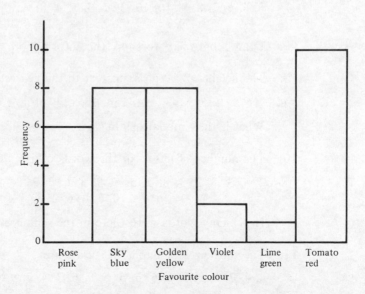

It shows that the most popular colour, or the modal colour, is tomato red.

EXERCISE 23b What is the mode of each of the following sets of numbers:

1. 10, 8, 12, 14, 12, 10, 12, 8, 10, 12, 4

2. 3, 9, 7, 9, 5, 4, 8, 2, 4, 3, 5, 9

3. 1.2, 1.8, 1.9, 1.2, 1.8, 1.7, 1.4, 1.3, 1.8

4. 58, 56, 59, 62, 56, 63, 54, 53

5. 5.9, 5.6, 5.8, 5.7, 5.9, 5.9, 5.8, 5.7

6. 26.2, 26.8, 26.4, 26.7, 26.5, 26.4, 26.6, 26.5, 26.4

7. The table shows the number of goals scored by a football club last season.

Number of goals	0	1	2	3	4	5	6
Frequency	12	16	7	4	2	0	1

Draw a bar chart to show these results and find the modal score.

8. Given below are the marks out of 10 obtained by 30 girls in a history test.

8, 6, 5, 7, 8, 9, 10, 10, 3, 7, 3, 5, 4, 8, 7, 8, 10, 9, 8, 7, 10, 9, 9, 7, 5, 4, 8, 1, 9, 8

Draw a bar chart to show this information and find the mode.

9. The heights of 10 girls, correct to the nearest centimetre, are:

155, 148, 153, 154, 155, 149, 162, 154, 156, 155

What is their modal height?

10. The number of letters in the words of a sentence were:

2, 4, 3, 5, 2, 3, 8, 2, 5, 7, 9, 3, 6, 3, 7, 3, 4, 9, 2, 3, 8, 3, 5, 2, 10, 3, 4, 6, 2, 3, 4

How many words were there in the sentence? What is the mode?

11. The shoe sizes of pupils in a class are:

4, 4, 7, 6, 5, 5, 6, 6, 6, 4, 5, 8, 6, 7, 4, 7, 9, 6, 5, 7, 6, 7, 8, 6, 4, 4, 4, 5, 5, 7, 7, 7, 5, 8, 6, 5

How many pupils are there in the class?

What is the modal shoe size?

MEDIAN

The median value of a set of numbers is the value of the middle number when they have been placed in ascending (or descending) order.

Imagine nine children arranged in order of their height.

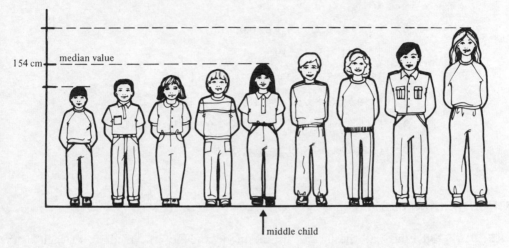

154 cm — median value

middle child

The height of the fifth or middle child is 154 cm,

i.e. the median height is 154 cm

Similarly 24 is the median of 12, 18, 24, 37 and 46. Two numbers are smaller than 24 and two are larger.

To find the median of 16, 49, 53, 8, 32, 19 and 62, rearrange the numbers in ascending order:

$$8, 16, 19, 32, 49, 53, 62$$

then we can see that the middle number of these is 32,

i.e. the median is 32.

If there is an even number of numbers, the median is found by finding the average or mean of the two middle values after they have been placed in ascending or descending order.

To find the median of 24, 32, 36, 29, 31, 34, 35, 39, rearrange in ascending order:

$$24, 29, 31, \underline{32}, \underline{34}, 35, 36, 39$$

Then the median is $\dfrac{32+34}{2} = \dfrac{66}{2}$

i.e. the median is 33.

EXERCISE 23c Find the median of each of the following sets of numbers.

1. 1, 2, 3, 5, 7, 11, 13

2. 26, 33, 39, 42, 64, 87, 90

3. 13, 24, 19, 13, 6, 36, 17

4. 4, 18, 32, 16, 9, 7, 29

5. 1.2, 3.4, 3.2, 6.5, 9.8, 0.4, 1.8

6. 5, 7, 11, 13, 17, 19

7. 34, 46, 88, 92, 104, 116, 118, 144

8. 34, 42, 16, 85, 97, 24, 18, 38

9. 1.92, 1.84, 1.89, 1.86, 1.96, 1.98, 1.73, 1.88

10. 15.2, 6.3, 14.8, 9.5, 16.3, 24.9

MIXED EXAMPLES

EXERCISE 23d Find   a) the mean   b) the mode and   c) the median, of each of the following sets of numbers:

1. 21, 16, 25, 21, 19, 32, 27

2. 67, 71, 69, 82, 70, 66, 81, 66, 67

3. 43, 46, 47, 45, 45, 42, 47, 49, 43, 43

4. 84, 93, 13, 16, 28, 13, 32, 63, 45

5. 30, 27, 32, 27, 28, 27, 26, 27

6. In seven rounds of golf, a golfer returns scores of: 72, 87, 73, 72, 86, 72 and 77. Find the mean, mode and median of these scores.

7. The heights (correct to the nearest centimetre) of a group of girls are: 159, 155, 153, 154, 157, 162, 152, 160, 161, 157. Find   a) their mean height   b) their modal height   c) their median height.

8. The marks, out of 100, in a geography test for the members of a class were: 64, 50, 35, 85, 52, 47, 72, 31, 74, 49, 36, 44, 54, 48, 32, 52, 53, 48, 71, 52, 56, 49, 81, 45, 52, 80, 46. Find   a) the mean mark   b) the modal mark   c) the median mark.

9. Find the mean, mode and median of the following golf scores:
85, 76, 91, 83, 88, 84, 84, 82, 77, 79, 80, 83, 86, 84.

10. The table shows how many pupils in a form were absent for various numbers of sessions during a certain school week.

Number of sessions absent	0	1	2	3	4	5	6	7	8	9	10
Frequency	20	2	4	0	2	0	1	2	0	0	1

Find   a) the mode   b) the median   c) the mean.

11. The table shows the number of children per family in the families of the pupils in a class.

Number of children	1	2	3	4	5	6	7
Frequency	1	3	9	5	5	2	1

Find   a) the mode   b) the median   c) the mean.

24 TRAVEL GRAPHS

FINDING DISTANCE FROM A GRAPH

When we went on holiday in the car we travelled to our holiday resort at a steady speed of 30 kilometres per hour (km/h), i.e. in each hour we covered a distance of 30 km.

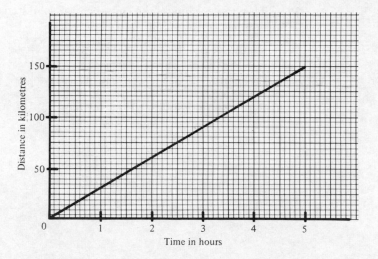

This graph shows our journey. It plots distance against time and shows that

in 1 hour we travelled   30 km
in 2 hours we travelled   60 km
in 3 hours we travelled   90 km
in 4 hours we travelled 120 km
in 5 hours we travelled 150 km

EXERCISE 24a The graphs that follow show ten different journeys. For each journey find:

a) the distance travelled

b) the time taken

c) the distance travelled: in 1 hour (questions 1, 2, 3, 6, 7 and 8)
or in 1 second (questions 4, 5, 9 and 10)

340

1.

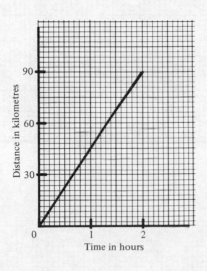

3.

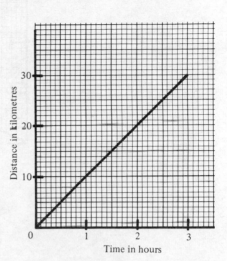

2.

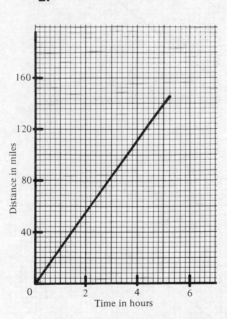

4.

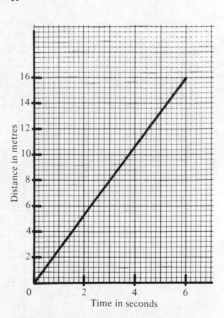

5.

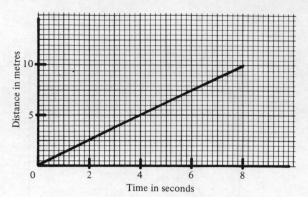

6.

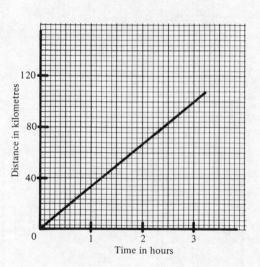

7.

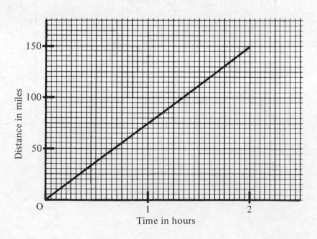

8.

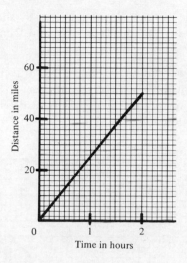

9.

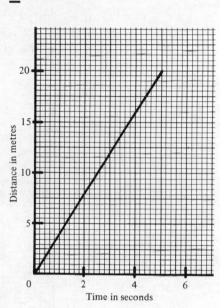

10.

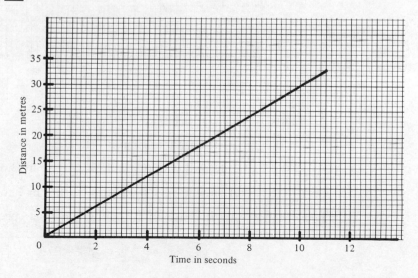

DRAWING TRAVEL GRAPHS

If Peter walks at 6 km/h, we can draw a graph to show this, using 2 cm to represent 12 km on the distance axis and 2 cm to represent 1 hour on the time axis.

Plot the point which shows that in 1 hour he has travelled 6 km. Join the origin to this point and produce the straight line to give the graph shown. From this graph we can see that in 2 hours Peter travels 12 km and in 5 hours he travels 30 km.

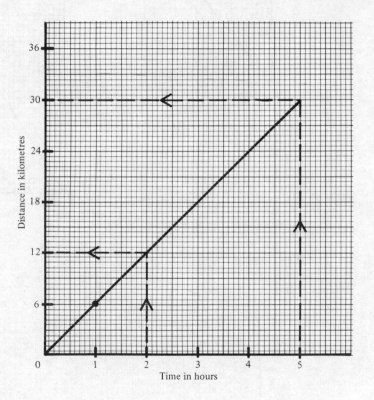

Alternatively we could say that

if he walks 6 km in 1 hour

he will walk 6×2 km $= 12$ km in 2 hours

and he will walk 6×5 km $= 30$ km in 5 hours

The distance walked is found by multiplying the speed by the time,

i.e.

$$\text{Distance} = \text{speed} \times \text{time}$$

EXERCISE 24b

Draw a travel graph to show a journey of 150 km in 3 hours. Plot distance along the vertical axis and time along the horizontal axis.

Let 4 cm represent 1 hour and 2 cm represent 50 km.

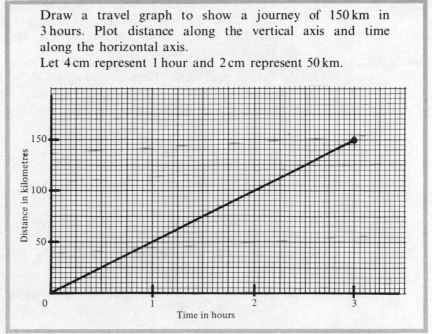

Draw travel graphs to show the following journeys. Plot distance along the vertical axis and time along the horizontal axis. Use the scales given in brackets.

1. 60 km   in 2 hours   (4 cm ≡ 1 hour,   1 cm ≡ 10 km)

2. 180 km   in 3 hours   (4 cm ≡ 1 hour,   2 cm ≡ 50 km)

3. 300 km   in 6 hours   (1 cm ≡ 1 hour,   1 cm ≡ 50 km)

4. 80 miles   in 2 hours   (6 cm ≡ 1 hour,   1 cm ≡ 10 miles)

5. 140 miles in 4 hours   (2 cm ≡ 1 hour,   1 cm ≡ 25 miles)

6. 100 km   in $2\frac{1}{2}$ hours (2 cm ≡ 1 hour,   2 cm ≡ 25 km)

7. 105 km   in $3\frac{1}{2}$ hours (2 cm ≡ 1 hour,   4 cm ≡ 50 km)

8. 75 miles   in $1\frac{1}{4}$ hours (8 cm ≡ 1 hour,   2 cm ≡ 25 miles)

9. 40 m   in 5 sec   (2 cm ≡ 1 sec,   2 cm ≡ 10 m)

10. 240 m   in 12 sec   (1 cm ≡ 1 sec,   2 cm ≡ 50 m)

11. Alan walks at 5 km/h. Draw a graph to show him walking for 3 hours. Take 4 cm to represent 5 km and 4 cm to represent 1 hour. Use your graph to find how far he walks in
a) $1\frac{1}{2}$ hours   b) $2\frac{1}{4}$ hours.

12. Julie can jog at 10 km/h. Draw a graph to show her jogging for 2 hours. Take 1 cm to represent 2 km and 8 cm to represent 1 hour. Use your graph to find how far she jogs in
a) $\frac{3}{4}$ hour   b) $1\frac{1}{4}$ hours.

13. Jo drives at 35 mph. Draw a graph to show her driving for 4 hours. Take 1 cm to represent 10 miles and 4 cm to represent 1 hour. Use your graph to find how far she drives in
a) 3 hours   b) $1\frac{1}{4}$ hours.

14. John walks at 4 mph. Draw a graph to show him walking for 3 hours. Take 1 cm to represent 1 mph and 4 cm to represent 1 hour. Use your graph to find how far he walks in
a) $\frac{1}{2}$ hour   b) $3\frac{1}{2}$ hours.

The remaining questions should be solved by calculation.

15. An express train travels at 200 km/h. How far will it travel in
a) 4 hours   b) $5\frac{1}{2}$ hours?

16. Ken cycles at 24 km/h. How far will he travel in
a) 2 hours   b) $3\frac{1}{2}$ hours   c) $2\frac{1}{4}$ hours?

17. An aeroplane flies at 300 mph. How far will it travel in
a) 4 hours   b) $5\frac{1}{2}$ hours?

18. A bus travels at 60 km/h. How far will it travel in
a) $1\frac{1}{2}$ hours   b) $2\frac{1}{4}$ hours?

19. Susan can cycle at 12 mph. How far will she ride in
a) $\frac{3}{4}$ hour   b) $1\frac{1}{4}$ hours?

20. An athlete can run at 10.5 m/s. How far will he travel in
a) 5 sec   b) 8.5 sec?

21. A boy cycles at 12 mph. How far will he travel in
a) 2 hours 40 min   b) 3 hours 10 min?

22. Majid can walk at 8 km/h. How far will he walk in
a) 30 min   b) 20 min   c) 1 hour 15 min?

23. A racing car travels at 111 mph. How far will it travel in
a) 20 min   b) 1 hour 40 min?

24. A bullet travels at 100 m/s. How far will it travel in
a) 5 sec   b) $8\frac{1}{2}$ sec?

25. A Boeing 747 travels at 540 mph. How far does it travel in
a) 3 hours 15 min   b) 7 hours 45 min?

26. A racing car travels around a 2 km circuit at 120 km/h. How many laps will it complete in   a) 30 min   b) 1 hour 12 min?

CALCULATING THE TIME TAKEN

Georgina walks at 6 km/h so we can find how long it will take her to walk   a) 24 km   b) 15 km.

a) If she takes 1 hour to walk 6 km,
   she will take $\frac{24}{6}$ hours, i.e. 4 h, to walk 24 km.

b) If she takes 1 hour to walk 6 km,
   she will take $\frac{15}{6}$ hours, i.e. $2\frac{1}{2}$ hours, to walk 15 km.

i.e.
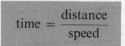
$$\text{time} = \frac{\text{distance}}{\text{speed}}$$

EXERCISE 24c   **1.** How long will Zena, walking at 5 km/h, take to walk
a) 10 km   b) 15 km?

2. How long will a car travelling at 80 km/h, take to travel
a) 400 km   b) 260 km?

3. How long will it take David, running at 10 mph, to run
a) 5 miles   b) $12\frac{1}{2}$ miles?

4. How long will it take an aeroplane flying at 450 mph to fly
a) 1125 miles   b) 2400 miles?

5. A cowboy rides at 14 km/h. How long will it take him to ride
a) 21 km   b) 70 km?

6. A rally driver drives at 50 mph. How long does it take him to
travel   a) 75 miles   b) 225 miles?

7. An athlete runs at 8 m/s. How long does it take him to cover
a) 200 m   b) 1600 m?

8. A dog runs at 20 km/h. How long will it take him to travel
a) 8 km   b) 18 km?

9. A liner cruises at 28 nautical miles per hour. How long will it
take to travel   a) 6048 nautical miles   b) 3528 nautical miles?

10. A car travels at 56 mph. How long does it take to travel
a) 70 miles   b) 154 miles?

11. A cyclist cycles at 12 mph. How long will it take him to cycle
a) 30 miles   b) 64 miles?

12. How long will it take a car travelling at 64 km/h to travel
a) 48 km   b) 208 km?

AVERAGE SPEED

Russell Compton left home at 8 a.m. to travel the 50 km to his place of work. He arrived at 9 a.m. Although he had travelled at many different speeds during his journey he covered the 50 km in exactly 1 hour. We say that his *average speed* for the journey was 50 kilometres per hour, or 50 km/h. If he had travelled at the same speed all the time, he would have travelled at 50 km/h.

Judy Smith travelled the 135 miles from her home to London in 3 hours. If she had travelled at the same speed all the time, she would have travelled at $\frac{135}{3}$ mph, i.e. 45 mph. We say that her average speed for the journey was 45 mph.

In each case:
$$\text{average speed} = \frac{\text{distance travelled}}{\text{time taken}}$$

This formula can also be written:

$$\text{distance travelled} = \text{average speed} \times \text{time taken}$$

and
$$\text{time taken} = \frac{\text{distance travelled}}{\text{average speed}}$$

Suppose that a car travels 35 km in 30 min, and we wish to find its speed in kilometres per hour. To do this we must express the time taken in hours instead of minutes,

i.e.
$$\text{time taken} = 30 \text{ min} = \tfrac{1}{2} \text{ hour}$$

Then
$$\text{average speed} = \frac{35}{\frac{1}{2}} \text{ km/h} = 35 \times \frac{2}{1} \text{ km/h}$$
$$= 70 \text{ km/h}$$

Great care must be taken with units. If we want a speed in kilometres per hour, we need the distance in kilometres and the time in hours. If we want a speed in metres per second, we need the distance in metres and the time in seconds.

EXERCISE 24d Find the average speed for each of the following journeys:

1.	80 km in 1 hour	**7.**	150 km in 3 hours
2.	120 km in 2 hours	**8.**	520 km in 8 hours
3.	60 miles in 1 hour	**9.**	245 miles in 7 hours
4.	480 miles in 4 hours	**10.**	104 miles in 13 hours
5.	80 m in 4 sec	**11.**	252 m in 7 sec
6.	135 m in 3 sec	**12.**	255 m in 15 sec

Find the average speed in km/h for a journey of 39 km which takes 45 min.

First, convert the time taken to hours:

$$45 \text{ min} = \frac{45}{60} \text{ hour} = \frac{3}{4} \text{ hour}$$

Then   average speed $= \dfrac{\text{distance travelled}}{\text{time taken}}$

$$= \frac{39 \text{ km}}{\frac{3}{4} \text{ hour}}$$

$$= 39 \times \frac{4}{3} \text{ km/h}$$

$$= 52 \text{ km/h}$$

Find the average speed in km/h for a journey of:

13. 40 km in 30 min   **15.** 48 km in 45 min

14. 60 km in 40 min   **16.** 66 km in 33 min

Find the average speed in km/h for a journey of 5000 m in $\frac{1}{2}$ hour.

$$5000 \text{ m} = \frac{5000}{1000} \text{ km} = 5 \text{ km}$$

average speed $= \dfrac{\text{distance travelled}}{\text{time taken}}$

$$= \frac{5 \text{ km}}{\frac{1}{2} \text{ hour}}$$

$$= 5 \times \frac{2}{1} \text{ km/h}$$

$$= 10 \text{ km/h}$$

Find the average speed in km/h for a journey of:

17. 4000 m in 20 min   **19.** 40 m in 8 sec

18. 6000 m in 45 min   **20.** 175 m in 35 sec

Find the average speed in mph for a journey of:

21. 27 miles in 30 min   **23.** 25 miles in 25 min

22. 18 miles in 20 min   **24.** 28 miles in 16 min

The following table shows the distances in kilometres between various places in the United Kingdom.

	London	Bradford	Cardiff	Leicester	Manchester	Oxford	Reading
Bradford	320						
Cardiff	250	332					
Leicester	160	160	224				
Manchester	310	55	277	138			
Oxford	90	280	172	120	230		
Reading	64	320	192	164	264	45	
York	315	53	390	174	103	290	210

Use this table to find the average speeds for journeys between:

25. London, leaving at 1025, and Manchester, arriving at 1625

26. Oxford, leaving at 0330, and Cardiff, arriving at 0730

27. Leicester, leaving at 1914, and Oxford, arriving at 2044

28. Reading, leaving at 0620, and London, arriving at 0750

29. Bradford, leaving at 1537, and Oxford, arriving at 1907

30. Cardiff, leaving at 1204, and York, arriving at 1624

31. Bradford, leaving at 1014, and Reading, arriving at 1638.

Problems frequently occur where different parts of a journey are travelled at different speeds in different times but we wish to find the average speed for the whole journey.

Consider for example a motorist who travels the first 50 miles of a journey at an average speed of 25 mph and the next 90 miles at an average speed of 30 mph.

One way to find his average speed for the whole journey is to complete the following table by using the relationship:

$$\text{time in hours} = \frac{\text{distance in miles}}{\text{speed in mph}}$$

	Speed in mph	Distance in miles	Time in hours
First part of journey	25	50	2
Second part of journey	30	90	3
Whole journey		**140**	**5**

We can add the distances to give the total length of the journey, and add the times to give the total time taken for the journey.

$$\text{average speed for whole journey} = \frac{\text{total distance}}{\text{total time}}$$

$$= \frac{140\,\text{miles}}{5\,\text{hours}}$$

$$= 28\,\text{mph}$$

Note: Never add or subtract average speeds.

We could also solve this problem, without using a table, as follows:

$$\text{time to travel 50 miles at 25 mph} = \frac{\text{distance}}{\text{speed}}$$

$$= \frac{50\,\text{miles}}{25\,\text{mph}}$$

$$= 2\,\text{hours}$$

$$\text{time to travel 90 miles at 30 mph} = \frac{\text{distance}}{\text{speed}}$$

$$= \frac{90\,\text{miles}}{30\,\text{mph}}$$

$$= 3\,\text{hours}$$

$\therefore$ total distance of 140 miles is travelled in 5 hours

i.e. $$\text{average speed for whole journey} = \frac{\text{total distance}}{\text{total time}}$$

$$= \frac{140\,\text{miles}}{5\,\text{hours}}$$

$$= 28\,\text{mph}$$

EXERCISE 24e
  1. I walk for 24 km at 8 km/h, and then jog for 12 km at 12 km/h. Find my average speed for the whole journey.

  2. A cyclist rides for 23 miles at an average speed of $11\frac{1}{2}$ mph before his cycle breaks down, forcing him to push his cycle the remaining distance of 2 miles at an average speed of 4 mph. Find his average speed for the whole journey.

  3. An athlete runs 6 miles at 8 mph, then walks 1 mile at 4 mph. Find his average speed for the total distance.

  4. A woman walks 3 miles at an average speed of $4\frac{1}{2}$ mph and then runs 4 miles at 12 mph. Find her average speed for the whole journey.

  5. A motorist travels the first 30 km of a journey at an average speed of 120 km/h, the next 60 km at 60 km/h, and the final 60 km at 80 km/h. Find the average speed for the whole journey.

  6. Phil Sharp walks the 2 km from his home to the bus stop in 15 min, and catches a bus immediately which takes him the 9 km to the railway station at an average speed of 36 km/h. He arrives at the station in time to catch the London train which takes him the 240 km to London at an average speed of 160 km/h. Calculate his average speed for the whole journey from home to London.

  7. A liner steaming at 24 knots takes 18 days to travel between two ports. By how much must it increase its speed to reduce the length of the voyage by 2 days?
(A knot is a speed of 1 nautical mile per hour.)

GETTING INFORMATION FROM TRAVEL GRAPHS

EXERCISE 24f

The graph opposite shows the journey of a coach which calls at three service stations A, B and C on a motorway. B is 60 km north of A and C is 20 km north of B. Use the graph to answer the following questions:

a) At what time does the coach leave A?

b) At what time does the coach arrive at C?

c) At what time does the coach pass B?

d) How long does the coach take to travel from A to C?

e) What is the average speed of the coach for the whole journey?

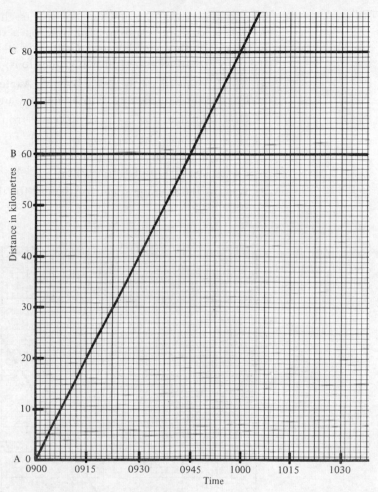

a) The coach leaves A at 0900.

b) It arrives at C at 1000.

c) It passes through B at 0945.

d) Time taken to travel from A to C is 1000 − 0900, i.e. 1 hour.

e) Distance from A to C = 60 km + 20 km = 80 km.

Time taken to travel from A to C = 1 hour.

$$\text{average speed} = \frac{\text{distance travelled}}{\text{time taken}}$$

$$= \frac{80\,\text{km}}{1\,\text{hour}} = 80\,\text{km/h}$$

1. The graph shows the journey of a car through three towns, Axeter, Bexley and Canton, which lie on a straight road. Axeter is 100 km south of Bexley and Canton is 60 km north of it. Use the graph to answer the following questions:

a) At what time does the car   i) leave Axeter
     ii) pass through Bexley   iii) arrive at Canton?

b) How long does the car take to travel from Axeter to Canton?

c) How long does the car take to travel
     i) the first 80 km of the journey?
     ii) the last 80 km of the journey?

d) What is the average speed of the car for the whole journey?

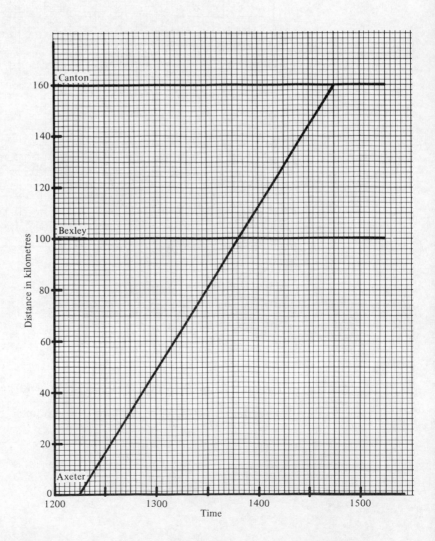

2. The graph shows the journey of an athlete in a race.

a) What was the length of the race?

b) How long did the athlete take?

c) What was his average speed for the whole journey?

d) How far did he travel in the first $1\frac{1}{4}$ hours?

e) Did the athlete stop at any time during the race?

f) Did the athlete travel at more than one speed?

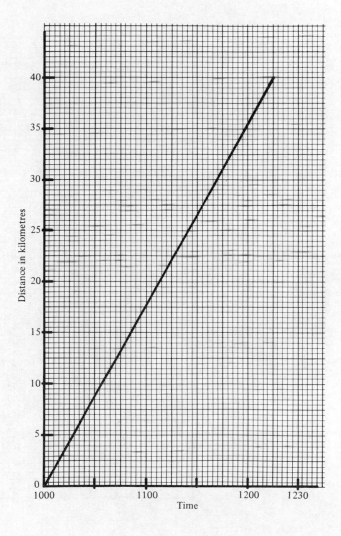

3. Sally went for a walk; the travel graph given below represents her journey.

a) How far did she walk?

b) At what time did she start?

c) How long did she take for the total distance?

d) What was her average speed?

e) How far did she walk in the first hour?

f) Did she walk at a constant speed?

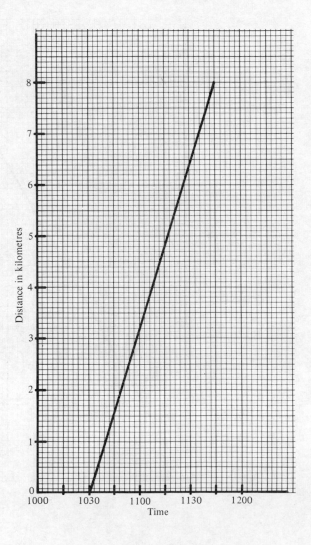

4. The graph shows the journey of an express train which starts
from A and passes through stations at B and C on the way to
its destination at D.

a) How far is it
   i) from A to B   ii) from B to C   iii) from C to D?

b) How long does the journey take
   i) from A to D   ii) from B to C?

c) Find the average speed for the whole journey.

d) Where is the train at 1100?

e) What time is it when the train is 20 miles short of C?

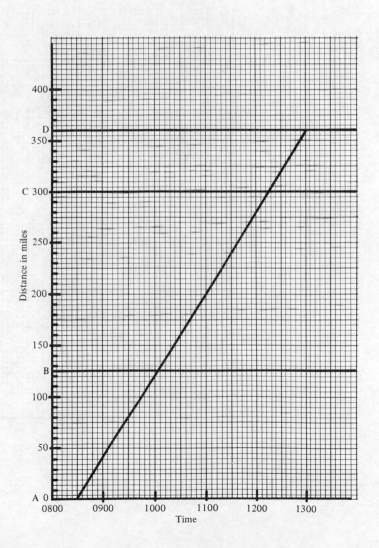

5. A coach leaves Newcombe at noon on its journey to Lee via Manley. The graph shows its journey.

a) How far is it
   i) from Newcombe to Manley   ii) from Manley to Lee?

b) How long does the coach take to travel from Newcombe to Lee?

c) What is the coach's average speed for the whole journey?

d) How far does the coach travel between 1.30 p.m. and 2.30 p.m.?

e) How far is the coach from
   i) Newcombe   ii) Manley, after travelling for $1\frac{1}{2}$ hours?

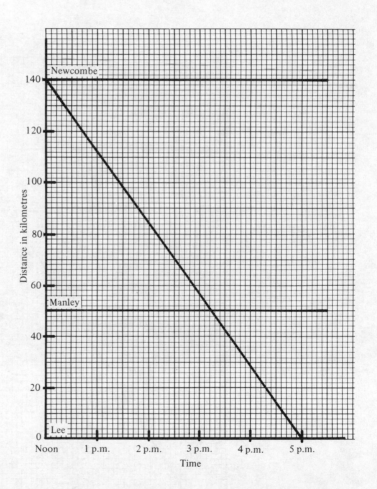

6. A cyclist leaves the seaside to cycle home. The graph shows his journey.
- a) At what time does he
  - i) leave the seaside   ii) arrive at home?
- b) How far is it from the seaside to his home?
- c) What is the average speed at which he cycles home?
- d) How long does he take to travel the first 10 miles?
- e) How far is he from home at 1430?
- f) What time is it when the cyclist has travelled 15 miles?

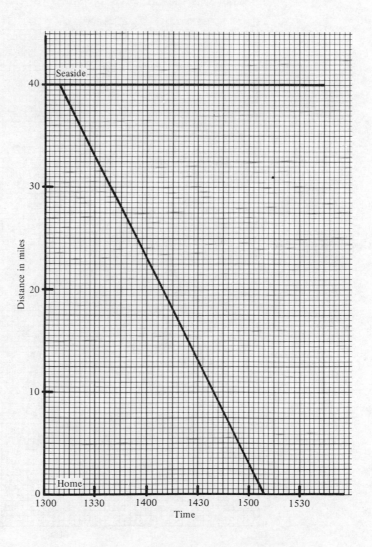

7. Father used the family car to transport the children from their home to the nearest mainline railway station and then returned home. The graph shows the journey.

a) How far is it from home to the station?

b) How long did it take the family to get to the station?

c) What was the average speed of the car on the journey to the station?

d) How long did the car take for the return journey?

e) What was the average speed for the return journey?

f) What was the car's average speed for the round trip?

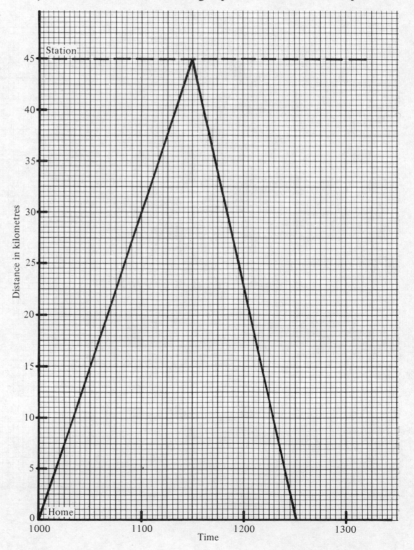

8. An athlete runs a certain distance, stops, turns around and walks back to her starting point. The graph shows her journey.

a) How far does she run?

b) For how long is she running?

c) What is her average running speed?

d) How far does she walk?

e) For how long is she walking?

f) What is her average walking speed?

g) What is her average speed for the whole journey?

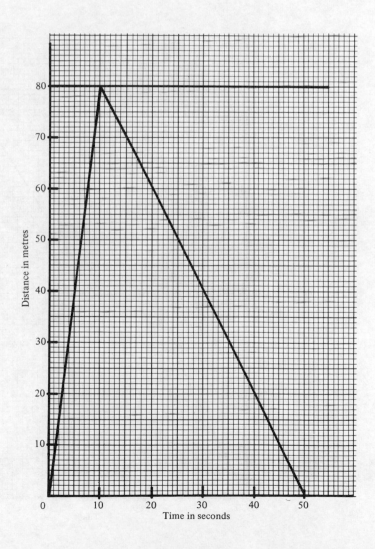

9. The graph shows the journey of a car through three service stations A, B and C, on a motorway.

a) Where was the car at   i) 0900   ii) 0930?

b) What was the average speed of the car between
   i) A and B   ii) B and C?

c) For how long does the car stop at B?

d) How long did the journey take?

e) What was the average speed of the car for the whole journey? Give your answer correct to 1 s.f.

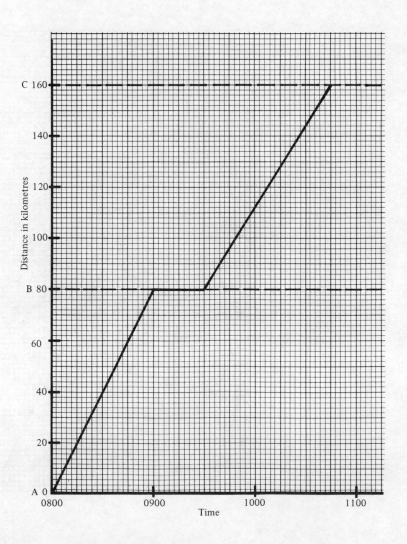

10. The graph shows Bill's journey on a sponsored walk.
  a) How far did he walk?
  b) How many times did he stop?
  c) What was the total time he spent resting?
  d) How long did he actually spend walking?
  e) How long did the walk take him?
  f) What was his average speed for the whole journey?
  g) Over which of the four stages did he walk fastest?
  h) Over which two stages did he walk at the same speed?

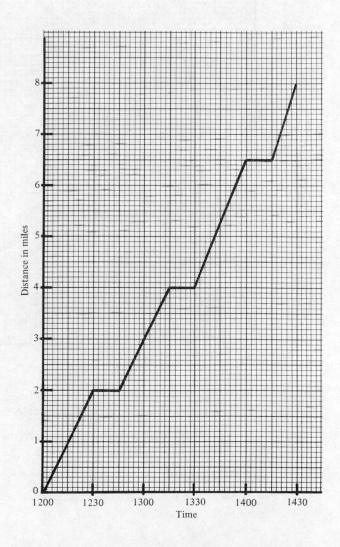

EXERCISE 24g

The graph shows Mrs Webb's journey on a bicycle to go shopping in the nearest town. Use it to answer the following questions:

a) How far is town from home?

b) How long did she take to get to town?

c) How long did she spend in town?

d) At what time did she leave for home?

e) What was her average speed on the outward journey?

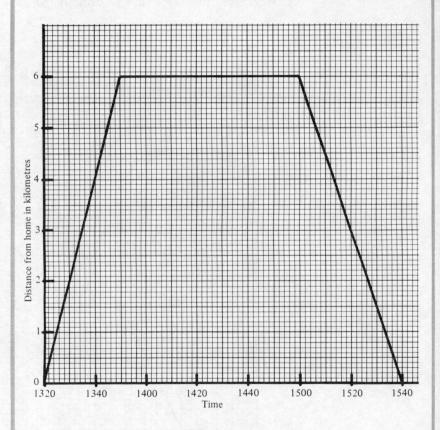

a) The graph shows that it is 6 km from home to town.

b) Mrs Webb left home at 1320 and arrived in town at 1350. The journey therefore took 30 min.

c) She arrived in town at 1350 and left at 1500. She therefore spent 1 hour 10 min there.

d) Mrs Webb left for home at 1500.

e) On the outward journey:

$$\text{Average speed} = \frac{\text{distance travelled}}{\text{time taken}}$$

$$= \frac{6\,\text{km}}{30\,\text{min}}$$

$$= \frac{6\,\text{km}}{\frac{1}{2}\,\text{hour}}$$

$$= 6 \times \frac{2}{1}\,\text{km/h}$$

$$= 12\,\text{km/h}$$

1. The graph shows the journey of a train from Newpool to London and back again. Use the graph to answer the questions that follow:

a) How far is Newpool from London?

b) How long did the outward journey take?

c) What was the average speed for the outward journey?

d) How long did the train remain in London?

e) At what time did the train leave London, and how long did the return journey take?

f) What was the average speed on the return journey?

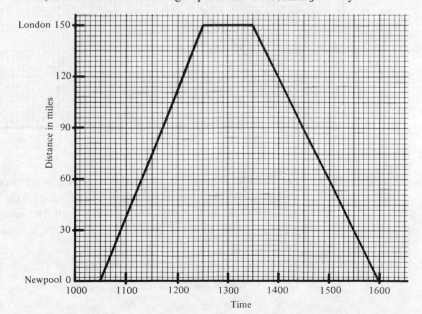

2. John Bloomfield's journey from A to C via B is shown on the graph.

a) At what time did he
   i) leave A   ii) arrive at B   iii) leave B   iv) arrive at C?

b) How far is it from A to C via B?

c) What was his average speed
   i) from A to B   ii) from B to C?

d) How long did he rest at B?

e) What was his average speed (including the stop) from A to C?

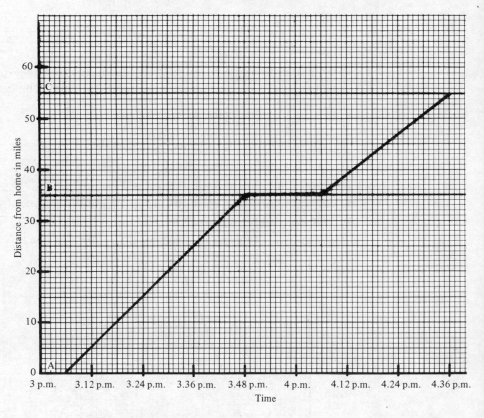

3. Opposite is the travel graph for two motorists travelling between London and Manley which are 110 miles apart. The first leaves Manley at 0900 for London, having a short break en route. The second leaves London at 1015 and travels non-stop to Manley. Use your graph to find

a) the average speed of each motorist for the complete journey,

b) when and where they pass,

c) their distance apart at 1200.

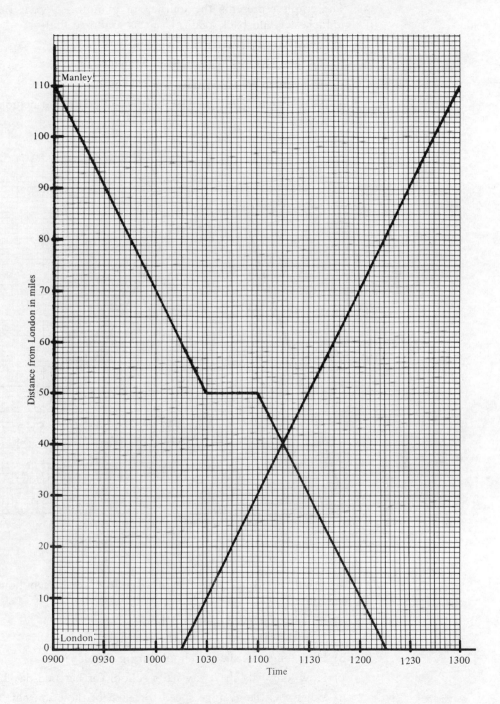

4. The graph represents the journey of a motorist from Leeds to Manchester and back again. Use this graph to find

a) the distance between the two cities,

b) the time the motorist spent in Manchester,

c) his average speed on the outward journey,

d) the average speed on the homeward journey (including the stop).

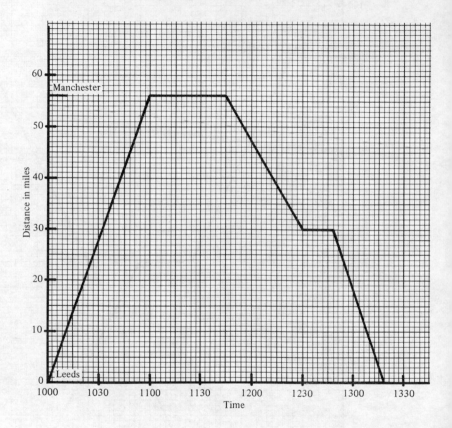

5. The graph opposite shows Judith's journeys between home and school.

a) At what time did she leave home
   i) in the morning   ii) in the afternoon?

b) How long was she in school during the day?

c) How long was she away from school for her mid-day break?

d) What was the average speed for each of these journeys?

e) Find the total time for which she was away from home.

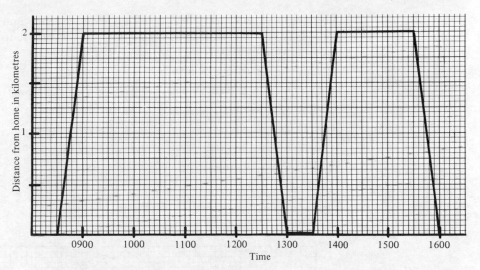

6. The graph below shows the journeys of two cars between two service stations, A and B, which are 180 km apart. Use the graph to find

a) the average speed of the first motorist and his time of arrival at B,

b) the average speed of the second motorist and the time at which she leaves B,

c) when and where the two motorists pass,

d) their distance apart at 1427.

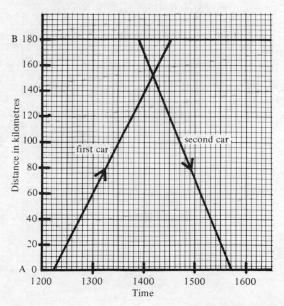

7. The graph shows the journeys of an express train and a slow train between two stations, X and Y. Use the graph to answer the following questions:

a) How far apart are X and Y?

b) Which train leaves Y—the express train or the slow train?

c) Find the average speed of each train.

d) Which train arrives at its destination first? How much longer is it before the other train arrives at its destination?

e) When and where do the two trains pass?

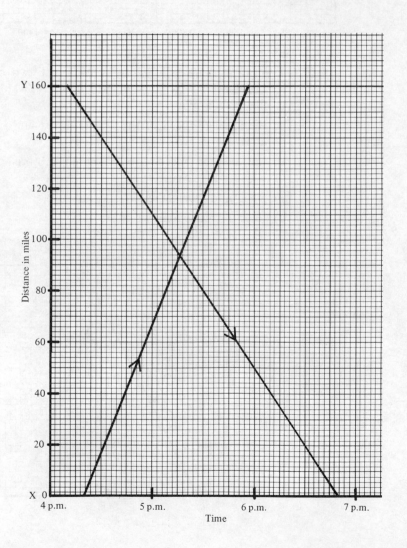

8. The graph represents the bicycle journeys of three school friends, Audrey, Betty and Chris, from the village in which they live to Buckwell, the nearest main town, which is 30 km away. Use the graph to find:

a) their order of arrival at Buckwell,

b) Audrey's average speed for the journey,

c) Betty's average speed for the journey,

d) Chris's average speed for the journey,

e) where and when Chris passes Audrey,

f) how far each is from town at 2 pm,

g) how far Betty is ahead of Chris at 2.15 pm.

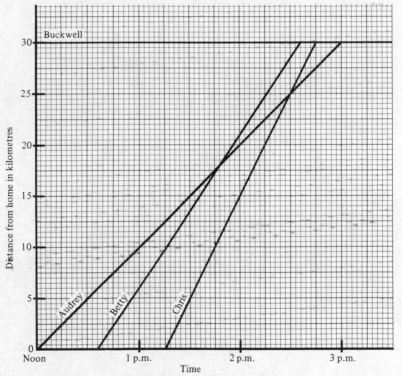

9. Jane leaves home at 1 p.m. to walk at a steady 4 mph towards Cornforth, which is 16 miles away, to meet her boyfriend Tim. Tim leaves Cornforth at 2.18 p.m. and jogs at a steady 6 mph to meet her. Draw a graph for each of these journeys taking 4 cm ≡ 1 hour on the time axis and 1 cm ≡ 1 mile on the distance axis. From your graph find:

a) when and where they meet,

b) their distance apart at 3 p.m.

10. A and B are motorway service areas 110 miles apart. A car leaves A at 2.16 p.m. and travels at a steady 63 mph towards B while a motorcycle leaves B at 2.08 p.m. and travels towards A at a steady 45 mph. Draw a graph for the journeys taking 6 cm ≡ 1 hour and 1 cm ≡ 5 miles. From your graph find:

a) when and where they pass,

b) where the motorcycle is when the car starts,

c) where the motorcycle is when the car arrives at B.

MIXED EXERCISES

EXERCISE 24h **1.** The graph shows John's walk from home to his grandparents' home.

a) How far away do they live?

b) How long did the journey take him?

c) What was his average walking speed?

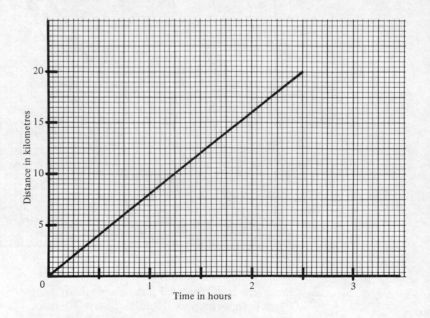

2. Jenny runs at 20 km/h. Draw a graph to show her running for $2\frac{1}{2}$ hours. Use your graph to find

a) how far she has travelled in $1\frac{3}{4}$ hours,

b) how long she takes to run the first 25 km.

3. A ship travels at 18 nautical miles per hour. How long will it take to travel   a) 252 nautical miles   b) 1026 nautical miles?

4. Find the average speed in km/h of a journey of 48 km in 36 min.

5. I left London at 1147 to travel the 315 miles to York. If I arrived at 1617, what was my average speed?

6. I walk $\frac{2}{3}$ mile in 10 min and then run $\frac{1}{3}$ mile in 2 min. What is my average speed for the whole journey?

7. The graph shows Paul's journey in a sponsored walk from A to B. On the way his sister, who is travelling by car in the opposite direction from B to A, passes him.

a) How far does Paul walk?

b) How long does he take?

c) How much of this time does he spend resting?

d) What is his average speed for the whole journey?

e) What is his sister's average speed?

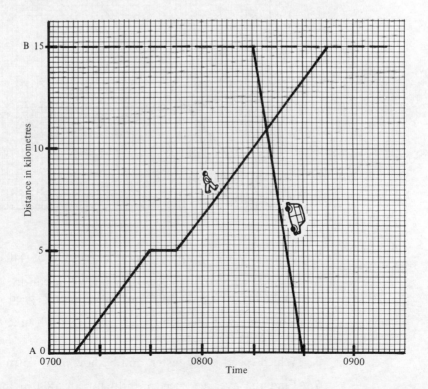

EXERCISE 24i   **1.**   The graph shows the journey of a scheduled non-stop express train from my home city to London.

   a) How far is my home city from London?

   b) How long did the journey take?

   c) What happened during the journey that was not intended?

   d) What was the average speed of the train for the first part of the journey?

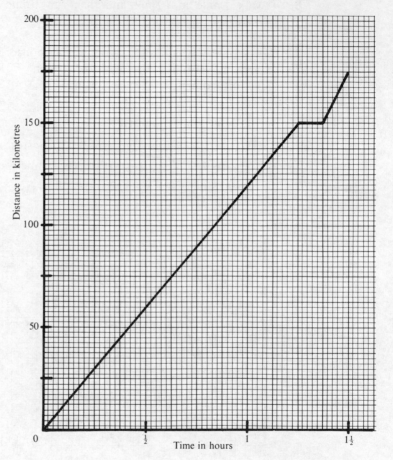

2.   Draw a travel graph to show a journey of 440 km in 4 hours.

3.   A horse runs at 15 m/sec. How far will it run in
   a) 1 min   b) $1\frac{3}{4}$ min?   Express its running speed in km/h.

4.   How long will a coach travelling at 72 km/h take to travel
   a) 216 km   b) 126 km.

5.   Which speed is the faster, and by how much: 50 m/sec or 200 km/h?

6.   Find the average speed (km/h) for an 1800 m journey in 9 min.

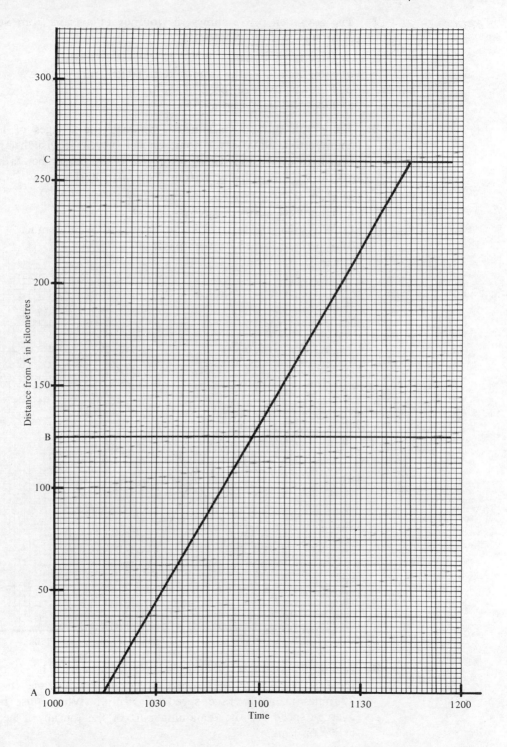

7. The graph on p.375 shows the journey of a train from Station A to Station C via Station B.

   a) How far is A from   i) B   ii) C?
   b) How long does the train take to travel from A to C?
   c) Find the average speed of the train.
   d) Does the train stop at B?

8. The graph shows two journeys between the villages of Farley and Weston. Nina leaves home on her bicycle to visit her friend who lives at Weston. On her way there she passes her father in his car who is on his way from Weston to Farley.

   a) How far is it between the two places?
   b) How long does each journey take?
   c) Which has the faster average speed and by how much?
   d) Where and when do they pass?

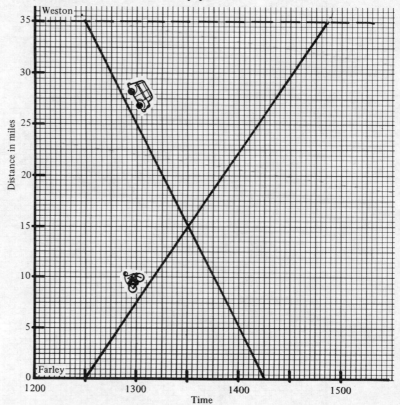

9. A motorist wants to make a 300 mile journey in $5\frac{1}{2}$ h. He travels the first 60 miles at an average speed of 45 mph, and the next 200 miles at an average speed of 60 mph. What must be his average speed for the remaining part of the journey if he is to arrive on time?

25 BILLS AND WAGES

SHOPPING BILLS

EXERCISE 25a Use your calculator to total the following supermarket bills. In each case find the change from a £20 note.

	£		£		£		£		£
1.	.88	**2.**	.62	**3.**	.55	**4.**	.36	**5.**	1.26
	.82		.37		.43		.72		.49
	.44		.37		.43		.42		.53
	.17		.37		.27		.42		.75
	.38		.42		.64		.93		.44
	.24		.18		.59		.45		.45
	.29		.23		.19		.45		.45
	.33		1.04		.19		.37		.45
	.34		.77		.54		.37		.45
	.23		.64		.62		.85		.62
	1.29		.53		.73		4.21		.41
	.29		.22		.80		.62		.87
	.59		.22		.34		.14		.73
	.43		.22		.37		.14		.49
	.23		.89		.52		.25		.61
	.32		.73		.49		.25		.72
	.32		.32		.26		.72		.17
	.28		.32		.37		.64		.17
	.16		2.76		1.04		.45		.43
	.77		3.49		.92		.27		.56
	1.43		.23		.76		.27		.92
	.49		.42		.43		.84		.44
	.42				.52		.92		.73
	.18						.66		.84
									.44
									.62

Copy and complete the following bills:

£

6. 2 tins of paint at £4.20 per tin
7 rolls of wallpaper at £5.20 per roll
3 brushes at £1.60 each

———————

£

7. 6 cakes at 24 p each
3 loaves of bread at 52 p each
1 currant loaf at 48 p

———————
———————

£

8. 2 kg butter at £2.20 per kilo
3 litres milk at 56 p per litre
2 cartons of cream at 98 p each

———————
———————

£

9. 2 packets of cereal at 87 p each
3 bags of flour at 54 p per bag
4 packets of soft brown sugar at 45 p per packet
7 packets of soup at 32 p per packet

———————
———————

£

10. 3 kg potatoes at 36 p per kilo
$1\frac{1}{2}$ kg carrots at 28 p per kilo
2 kg onions at 21 p per kilo

———————
———————

£

11. 7 oranges at 9 p each
8 grapefruit at 26 p each
3 lb apples at 32 p per lb
2 lb bananas at 38 p per lb

———————
———————

£

12. 16 lb potatoes at 19 p per lb
5 lb carrots at 30 p per lb
2 lb parsnips at 22 p per lb
3 lb beetroot at 24 p per lb

———————

£

13.   9 jellies at 14 p each
3 jars of jam at 62 p per jar
2 jars of marmalade at 84 p per jar
3 jars of honey at 92 p per jar

_____
_____

£

14.   3 bars of chocolate at 56 p per bar
7 packets of sweets at 45 p per packet
9 bags of crisps at 17 p per bag

_____
_____

£

15.   7 newspapers at 25 p each
2 magazines at 65 p each
3 comics at 18 p each
Delivery charge 24 p

_____
_____

£

16.   3 shirts at £12.50 each
2 ties at £3.25 each
6 pairs of socks at £2.30 per pair
1 pullover at £15.60

_____
_____

£

17.   3 skirts at £17.25 each
4 jumpers at £9.40 each
1 dress at £32.95
6 pairs of tights at 75 p per pair

_____
_____

£

18.   3 lb meat at £2.46 per lb
1½ lb bacon at £1.86 per lb
2 dozen eggs at 46 p per half dozen
5 packets of frozen mixed vegetables
at 88 p per packet

_____
_____

£

19. 2 demijohns at £1.56 each
3 cans of grapejuice at £3.45 per can
2 airlocks at 76 p each
1 packet of yeast at 57 p
1 packet of corks at 46 p

_____
_____

WAGES

Everybody who goes to work expects to get paid. Some are paid an annual amount or *salary*, but many people are paid a wage at a fixed sum per hour. There is usually an agreed length to the working week and any hours worked over and above this may be paid for at a higher rate.

If John Duffy works for 37 hours for an agreed hourly rate of £4.50, he receives payment of £4.50 × 37, i.e. £166.50. This figure is called his *gross wage* for the week. From this, deductions are made for such things as National Insurance contributions and Income Tax. After the deductions have been made he receives his *net wage* or "take-home" pay.

All this information is gathered together by the employer on a pay slip, an example of which is given below.

STAFF No.		DATE	Basic Salary	Additional Payts. A	Deduction for Absence	Gross Pay
01035932		JAN 1983	130.34	24.44		154.78
Attachments	Loan Repayts/ Adv. Recovered	Vol. Dedns. B	Nat. Ins.		Income Tax	Total Deducted
			13.54		46.50	60.04
A — Overtime	Commission	Bonuses	Other	Non-Taxble. Allces	NET PAY
24.44				16.78	111.52
Detail	Detail	Detail	Detail	B — Voluntary Deductions	

Taxable to date   377.71
Tax paid to date   113.10
Nat. Ins. to date   13.54
Pension (tax yr.)
Loan balances

EXERCISE 25b Calculate the gross weekly wage for each of the following factory workers.

	Name	Number of hours worked	Hourly rate of pay
1.	E. D. Nisbett	40	£3
2.	A. Tucker	35	£3.50
3.	D. A. Wilcox	38	£2.46
4.	H. J. Shore	39	£4.52
5.	T. Greenhalgh	$38\frac{1}{2}$	£3.86
6.	A. Smith	44	£4.46
7.	D. Thomas	$39\frac{1}{2}$	£5.58

In the questions that follow, it is assumed that the meal breaks are unpaid.

Sally Green works a five-day week Monday to Friday. She starts work every day at 8 a.m. and finishes at 4.30 p.m. She has 1 hour off for lunch. How many hours does she work in a week? Find her gross pay if her rate is £2.46 for each hour worked.

Number of hours from 8 a.m. to 4.30 p.m. is $8\frac{1}{2}$.

Since she has 1 hour off for lunch,

number of hours worked each day is $7\frac{1}{2}$

number of hours worked each week $= 7\frac{1}{2} \times 5$

$$= 37\frac{1}{2}$$

Gross pay for the week $= £2.46 \times 37\frac{1}{2}$

$$= £92.25$$

8. Edna Owen works a five-day week. She starts work each day at 7.30 a.m. and finishes at 4.15 p.m. She has 45 minutes for lunch and a 10 minute break each morning and afternoon. How long does she actually work a) in a day b) in a week? If her hourly rate is £2.66, calculate her gross wage for the week.

9. Martin Jones starts work each day at 7 a.m. and finishes at 4.30 p.m. He has a 45 minute lunch break. How many hours does he work in a normal five-day week? Find his gross weekly wage if his rate of pay is £3.24 per hour.

10. Jean Spann works "afternoons". She starts every day at 2 p.m. and finishes at 10.30 p.m., and is entitled to a meal break from 6 p.m. to 6.45 p.m. How many hours does she work a) in a day b) in a five-day week? Calculate her gross weekly wage if she is paid £2.26 per hour.

Mary Killick gets paid £2.14 per hour for her normal working week of $37\frac{1}{2}$ hours. Any overtime is paid at time-and-a-half. Find her gross pay in a week when she works $45\frac{1}{2}$ hours.

$$\text{Basic weekly pay} = £2.14 \times 37.5$$

$$= £80.25$$

$$\text{Number of hours overtime} = (45\tfrac{1}{2} - 37\tfrac{1}{2}) \text{ hours}$$

$$= 8 \text{ hours}$$

Since overtime is paid at time-and-a-half,

the rate of overtime pay is £2.14 × 1.5 = £3.21 per hour

$$\text{Payment for overtime} = £3.21 \times 8$$

$$= £25.68$$

$$\text{Total gross pay} = \text{Basic pay} + \text{Overtime pay}$$

$$= £80.25 + £25.68$$

$$= £105.93$$

11. Tom Shepherd works for a builder who pays £3.10 per hour for a basic week of 38 hours. If overtime worked is paid at time-and-a-half, how much will he earn in a week when he works for a) 38 hours b) 48 hours c) 50 hours?

12. Elsie Quinn works in a factory where the basic hourly rate is £1.96 for a 35 hour week. Any overtime is paid at time-and-a-half. How much will she earn in a week when she works for 46 hours?

13. Walter Markland works a basic week of $37\frac{1}{2}$ hours. Overtime is paid at time-and-a-quarter. How much does he earn in a week when he works $44\frac{1}{2}$ hours if the hourly rate is £3.40?

14. Peter Ambler's time sheet showed that he worked 7 hours overtime in addition to his basic 38 hour week. If his basic hourly rate is £3.16 and overtime is paid at time-and-a-half, find his gross pay for the week.

15. During a certain week Peggy Edwards worked $8\frac{1}{2}$ hours Monday to Friday together with 4 hours on Saturday. The normal working day was 7 hours and any time worked in excess of this was paid at time-and-a-half, with Saturday working being paid at double time. Calculate her gross wage for the week if she was paid £2.16 per hour.

16. Diana Read works a basic week of 39 hours. Overtime is paid at time-and-a-half. How much does she earn in a week when she works $47\frac{1}{2}$ hours if the hourly rate is £3.64?

17. Joan Danby's pay slip showed that she had worked $5\frac{1}{2}$ hours overtime in addition to her basic 37 hour week. If her basic rate of pay is £2.96 and overtime is paid at time-and-a-half, find her gross pay for the week.

18. Copy and complete the following table, which gives Norman Coleman's clocking in and clocking out times for a certain week.

Day	Morning		Afternoon		Hours worked
	Clocked in	Clocked out	Clocked in	Clocked out	
Monday	7.30 a.m.	12.15 p.m.	1.00 p.m.	4.15 p.m.	
Tuesday	7.30 a.m.	12.15 p.m.	1.00 p.m.	4.15 p.m.	
Wednesday	7.30 a.m.	12.15 p.m.	1.00 p.m.	4.15 p.m.	
Thursday	7.45 a.m.	12.15 p.m.	1.00 p.m.	4.15 p.m.	
Friday	7.30 a.m.	12.15 p.m.	1.00 p.m.	4.15 p.m.	
Saturday	7.30 a.m.	12 noon			

Norman Coleman's basic hourly rate is £2.88 and any hours worked in excess of 37 are paid at time-and-a-quarter. Calculate his gross wage for the week.

19. The timesheet for Anne Stent showed that during the last week in November she worked as follows:

Day	Morning		Afternoon	
	In	Out	In	Out
Monday	7.45 a.m.	12 noon	1.00 p.m.	5.45 p.m.
Tuesday	7.45 a.m.	12 noon	1.00 p.m.	4.15 p.m.
Wednesday	7.45 a.m.	12 noon	1.00 p.m.	4.15 p.m.
Thursday	7.45 a.m.	12 noon	1.00 p.m.	4.15 p.m.
Friday	7.45 a.m.	12 noon	1.00 p.m.	4.15 p.m.

a) What is the length of her normal working day?

b) How many hours make up her basic working week?

c) Calculate her basic weekly wage if the hourly rate is £2.84.

d) How much overtime was worked?

e) Calculate her gross wage if overtime is paid at time-and-a-half.

TELEPHONE BILLS

The cost of a telephone call depends on three factors:

i) the distance between the caller and the person being called,

ii) the time of day and/or the day of the week on which the call is being made,

iii) the length of the call.

These three factors are put together in various ways to give metered units of time, each unit being charged at a fixed rate.

In common with gas and electricity there is a standing charge each quarter in addition to the charge for the metered units.

For example, suppose that Chris Reynolds' telephone account for the last quarter showed that his telephone had been used for 546 metered units. If the standing charge was £15.50 and each unit cost 5 p, his telephone bill for the quarter can be worked out as follows:

$$\text{Cost of 546 units at 5 p per unit} = 546 \times 5\,\text{p}$$

$$= £27.30$$

$$\text{Standing charge} = £15.50$$

∴   the telephone bill for the quarter was £42.80

British TELECOM

BRITISH TELECOMMUNICATIONS

VAT registration no 243 1700 02

BRITISH TELECOM WEST MIDLAND
Service Manager's Office
Providence House
Charles Street
WORCESTER
WR1 2BE
Tel. Worcester (0905) 612311
Or ask operator for Freefone 4521
Telex: 335403 (BT WOR G)

BB FST R7

Any call charges
not to hand
when this bill
was prepared
will be included
in a later bill

See Notes Overleaf	Payment Is Now Due	Telephone number	Date of bill
			16 NOV 83 (Tax point)

Rental and other standing charges	from	to	£ quarterly rate	£
	1 NOV	31 JAN	19.40	19.40

Metered units (See overleaf)	date	meter reading	units used	
	11 AUG	025101		
	7 NOV	025715		
UNITS AT 4.30P			614	26.40

```
19 SEP   0.18  LOWER CHARGE
    CAREY    661                        7.43
TOTAL (EXCLUSIVE OF VAT)               53.23
VALUE ADDED TAX AT 15.00%               7.98
          TOTAL PAYABLE                61.21
```

For Office
use only

Paid
£

Initials

AX6060
R.P. LTD   PLEASE RETURN THE COUNTERFOIL BELOW WITH YOUR PAYMENT

EXERCISE 25c Find the quarterly telephone bill for each of the following households.

	Name	Number of units used	Standing charge	Cost per unit
1.	Mrs Keeling	750	£14	5 p
2.	Mr Hodge	872	£16	6 p
3.	Miss Hutton	1040	£16.50	7 p
4.	Mr Tucker	1213	£15.25	8 p
5.	Mrs Lings	957	£18.50	9 p
6.	Miss Jacob	1134	£18.80	8.5p
7.	Mr Higgins	765	£23.60	6.6 p
8.	Mrs Buckley	1590	£18.40	8.3 p
9.	Mr Leeson	765	£21	7.68 p
10.	Mrs Solly	965	£25.50	10.5 p
11.	Miss Tring	655	£17.60	8 p
12.	Mr White	764	£16.75	7.6 p
13.	Mrs Green	944	£19.30	8.2 p
14.	Mr Turner	1166	£20.15	9.75 p
15.	Miss Parker	1207	£17.95	7.88 p

ELECTRICITY: KILOWATT HOURS

We all use electricity in some form and we know that some appliances cost more to run than others. For example, an electric fire costs much more to run than a light bulb. Electricity is sold in units called kilowatt-hours (kWh) and each appliance has a rating that tells us how many kilowatt-hours it uses each hour.

A typical rating for an electric fire is 2 kW. This tells us that it will use 2 kWh each hour, i.e. 2 units per hour. On the other hand, a light bulb can have a rating of 100 W. Since 1 kilowatt = 1000 watts (kilo means "thousand" as we have already seen in kilometre and kilogram), the light bulb uses $\frac{1}{10}$ kWh each hour, or $\frac{1}{10}$ of a unit.

EXERCISE 25d How many units (i.e. kilowatt-hours) will each of the given appliances use in 1 hour?

1. a 3 kW electric fire	**7.** a 60 W video recorder
2. a 100 W bulb	**8.** a 20 W radio
3. a 1$\frac{1}{2}$ kW fire	**9.** an 8 kW cooker
4. a 60 W bulb	**10.** a 7 kW shower
5. a 1200 W hair dryer	**11.** a 145 W food mixer
6. a 250 W television set	**12.** a 2 kW dishwasher

With the help of an adult, find the rating of any of the following appliances that you might have at home. The easiest place to find this information is probably from the instructions.

13. an electric kettle	**17.** the television set
14. a shaver	**18.** a bedside lamp
15. the refrigerator	**19.** the main bulb in the living room
16. the washing machine	**20.** the electric cooker

How many units of electricity would

21. a 2 kW fire use in 8 hours

22. a 100 W bulb use in 10 hours

23. an 8 kW cooker use in 1$\frac{1}{2}$ hours

24. a 60 W bulb use in 50 hours

25. a 150 W refrigerator use in 12 hours

26. a 300 W television set use in 5 hours

27. a 12 W radio use in 12 hours

28. an 8 W night bulb use in a week at 10 hours per night

29. an 8 kW shower heater use in 15 min

30. a 5 W clock use in 1 week

For how long could the following appliances be run on one unit of electricity?

31. a 250 W bulb	**34.** a 100 W television set
32. a 2 kW electric fire	**35.** a 360 W electric drill
33. a 4 W radio	**36.** a 150 W food processor

In the following questions assume that 1 unit of electricity costs 6 p.

How much does it cost to run

37. a 100 W bulb for 5 hours

38. a 250 W television set for 8 hours

39. a 3 W clock for 1 week

40. a 3 kW kettle for 5 min

41. a 150 W refrigerator for 20 hours

ELECTRICITY BILLS

It is clear from the questions in the previous exercise that lighting from electricity is cheap but heating is expensive.

While electricity is a difficult form of energy to store, it is convenient to produce it continuously at the power stations, 24 hours a day. There are therefore times of the day when more electricity is produced than is normally required. The Electricity Boards are able to solve this problem by selling "off-peak", or "white meter", electricity to domestic users at a cheaper rate. Most of the electricity consumed in this way is for domestic heating.

Domestic electricity bills are calculated by charging every household a fixed amount, together with a charge for each unit used. Off-peak electricity is sold at approximately half price. The amount used is recorded on a meter, the difference between the readings at the beginning and end of a quarter showing how much has been used.

EXERCISE 25e

Mrs Comerford uses 1527 units of electricity in a quarter. If the standing charge is £9.45 and each unit costs 8 p, how much does electricity cost her for the quarter?

Cost of 1527 units at 8 p per unit $= 1527 \times 8$ p

$= £122.16$

Standing charge $= £9.45$

Total bill $= £131.61$

Find the quarterly electricity bills for each of the following households:

	Name	Number of units used	Standing charge	Cost per unit
1.	Mr George	500	£10	5 p
2.	Mrs Newton	600	£12	5 p
3.	Miss Ying	800	£15	8 p
4.	Mrs Kimber	1000	£10	9 p
5.	Mr Churchman	950	£15	10 p
6.	Mr Khan	750	£14	12 p
7.	Mrs Angel	1200	£20	10 p
8.	Mr Archer	756	£10.50	5 p
9.	Miss Deats	892	£12.50	9 p
10.	Mrs Posnett	1045	£9.75	7 p
11.	Mr Ryder	639	£18.30	8.2 p
12.	Mr Vincent	1427	£15.90	6.65 p
13.	Mrs Jackson	684	£18	11 p
14.	Mr Wilton	938	£16.40	7.36 p
15.	Mr Perry	1604	£13.75	8.94 p

Find the quarterly electricity bills for each of the following households. Assume in each case that there is a standing charge of £10, and that off-peak units are bought at half price.

	Name	Number of units used		Basic cost per unit
		At the basic price	Off-peak	
16.	Mr Bennett	1000	500	10 p
17.	Miss Cann	800	600	8 p
18.	Mrs Beaton	750	400	9 p
19.	Mr Hadley	640	1200	7.5 p
20.	Mrs Cummings	850	2500	8.2 p

NATURAL SINES

°	0′ 0.0°	6′ 0.1°	12′ 0.2°	18′ 0.3°	24′ 0.4°	30′ 0.5°	36′ 0.6°	42′ 0.7°	48′ 0.8°	54′ 0.9°
0	0.000	0.002	0.003	0.005	0.007	0.009	0.010	0.012	0.014	0.016
1	0.017	0.019	0.021	0.023	0.024	0.026	0.028	0.030	0.031	0.033
2	0.035	0.037	0.038	0.040	0.042	0.044	0.045	0.047	0.049	0.051
3	0.052	0.054	0.056	0.058	0.059	0.061	0.063	0.065	0.066	0.068
4	0.070	0.071	0.073	0.075	0.077	0.078	0.080	0.082	0.084	0.085
5	0.087	0.089	0.091	0.092	0.094	0.096	0.098	0.099	0.101	0.103
6	0.105	0.106	0.108	0.110	0.111	0.113	0.115	0.117	0.118	0.120
7	0.122	0.124	0.125	0.127	0.129	0.131	0.132	0.134	0.136	0.137
8	0.139	0.141	0.143	0.144	0.146	0.148	0.150	0.151	0.153	0.155
9	0.156	0.158	0.160	0.162	0.163	0.165	0.167	0.168	0.170	0.172
10	0.174	0.175	0.177	0.179	0.181	0.182	0.184	0.186	0.187	0.189
11	0.191	0.193	0.194	0.196	0.198	0.199	0.201	0.203	0.204	0.206
12	0.208	0.210	0.211	0.213	0.215	0.216	0.218	0.220	0.222	0.223
13	0.225	0.227	0.228	0.230	0.232	0.233	0.235	0.237	0.239	0.240
14	0.242	0.244	0.245	0.247	0.249	0.250	0.252	0.254	0.255	0.257
15	0.259	0.261	0.262	0.264	0.266	0.267	0.269	0.271	0.272	0.274
16	0.276	0.277	0.279	0.281	0.282	0.284	0.286	0.287	0.289	0.291
17	0.292	0.294	0.296	0.297	0.299	0.301	0.302	0.304	0.306	0.307
18	0.309	0.311	0.312	0.314	0.316	0.317	0.319	0.321	0.322	0.324
19	0.326	0.327	0.329	0.331	0.332	0.334	0.335	0.337	0.339	0.340
20	0.342	0.344	0.345	0.347	0.349	0.350	0.352	0.353	0.355	0.357
21	0.358	0.360	0.362	0.363	0.365	0.367	0.368	0.370	0.371	0.373
22	0.375	0.376	0.378	0.379	0.381	0.383	0.384	0.386	0.388	0.389
23	0.391	0.392	0.394	0.396	0.397	0.399	0.400	0.402	0.404	0.405
24	0.407	0.408	0.410	0.412	0.413	0.415	0.416	0.418	0.419	0.421
25	0.423	0.424	0.426	0.427	0.429	0.431	0.432	0.434	0.435	0.437
26	0.438	0.440	0.442	0.443	0.445	0.446	0.448	0.449	0.451	0.452
27	0.454	0.456	0.457	0.459	0.460	0.462	0.463	0.465	0.466	0.468
28	0.469	0.471	0.473	0.474	0.476	0.477	0.479	0.480	0.482	0.483
29	0.485	0.486	0.488	0.489	0.491	0.492	0.494	0.495	0.497	0.498
30	0.500	0.502	0.503	0.505	0.506	0.508	0.509	0.511	0.512	0.514
31	0.515	0.517	0.518	0.520	0.521	0.522	0.524	0.525	0.527	0.528
32	0.530	0.531	0.533	0.534	0.536	0.537	0.539	0.540	0.542	0.543
33	0.545	0.546	0.548	0.549	0.550	0.552	0.553	0.555	0.556	0.558
34	0.559	0.561	0.562	0.564	0.565	0.566	0.568	0.569	0.571	0.572
35	0.574	0.575	0.576	0.578	0.579	0.581	0.582	0.584	0.585	0.586
36	0.588	0.589	0.591	0.592	0.593	0.595	0.596	0.598	0.599	0.600
37	0.602	0.603	0.605	0.606	0.607	0.609	0.610	0.612	0.613	0.614
38	0.616	0.617	0.618	0.620	0.621	0.623	0.624	0.625	0.627	0.628
39	0.629	0.631	0.632	0.633	0.635	0.636	0.637	0.639	0.640	0.641
40	0.643	0.644	0.645	0.647	0.648	0.649	0.651	0.652	0.653	0.655
41	0.656	0.657	0.659	0.660	0.661	0.663	0.664	0.665	0.667	0.668
42	0.669	0.670	0.672	0.673	0.674	0.676	0.677	0.678	0.679	0.681
43	0.682	0.683	0.685	0.686	0.687	0.688	0.690	0.691	0.692	0.693
44	0.695	0.696	0.697	0.698	0.700	0.701	0.702	0.703	0.705	0.706

NATURAL SINES

°	0' 0.0°	6' 0.1°	12' 0.2°	18' 0.3°	24' 0.4°	30' 0.5°	36' 0.6°	42' 0.7°	48' 0.8°	54' 0.9°
45	0.707	0.708	0.710	0.711	0.712	0.713	0.714	0.716	0.717	0.718
46	0.719	0.721	0.722	0.723	0.724	0.725	0.727	0.728	0.729	0.730
47	0.731	0.733	0.734	0.735	0.736	0.737	0.738	0.740	0.741	0.742
48	0.743	0.744	0.745	0.747	0.748	0.749	0.750	0.751	0.752	0.754
49	0.755	0.756	0.757	0.758	0.759	0.760	0.762	0.763	0.764	0.765
50	0.766	0.767	0.768	0.769	0.771	0.772	0.773	0.774	0.775	0.776
51	0.777	0.778	0.779	0.780	0.782	0.783	0.784	0.785	0.786	0.787
52	0.788	0.789	0.790	0.791	0.792	0.793	0.794	0.795	0.797	0.798
53	0.799	0.800	0.801	0.802	0.803	0.804	0.805	0.806	0.807	0.808
54	0.809	0.810	0.811	0.812	0.813	0.814	0.815	0.816	0.817	0.818
55	0.819	0.820	0.821	0.822	0.823	0.824	0.825	0.826	0.827	0.828
56	0.829	0.830	0.831	0.832	0.833	0.834	0.835	0.836	0.837	0.838
57	0.839	0.840	0.841	0.842	0.842	0.843	0.844	0.845	0.846	0.847
58	0.848	0.849	0.850	0.851	0.852	0.853	0.854	0.854	0.855	0.856
59	0.857	0.858	0.859	0.860	0.861	0.862	0.863	0.863	0.864	0.865
60	0.866	0.867	0.868	0.869	0.869	0.870	0.871	0.872	0.873	0.874
61	0.875	0.875	0.876	0.877	0.878	0.879	0.880	0.880	0.881	0.882
62	0.883	0.884	0.885	0.885	0.886	0.887	0.888	0.889	0.889	0.890
63	0.891	0.892	0.893	0.893	0.894	0.895	0.896	0.896	0.897	0.898
64	0.899	0.900	0.900	0.901	0.902	0.903	0.903	0.904	0.905	0.906
65	0.906	0.907	0.908	0.909	0.909	0.910	0.911	0.911	0.912	0.913
66	0.914	0.914	0.915	0.916	0.916	0.917	0.918	0.918	0.919	0.920
67	0.921	0.921	0.922	0.923	0.923	0.924	0.925	0.925	0.926	0.927
68	0.927	0.928	0.928	0.929	0.930	0.930	0.931	0.932	0.932	0.933
69	0.934	0.934	0.935	0.935	0.936	0.937	0.937	0.938	0.938	0.939
70	0.940	0.940	0.941	0.941	0.942	0.943	0.943	0.944	0.944	0.945
71	0.946	0.946	0.947	0.947	0.948	0.948	0.949	0.949	0.950	0.951
72	0.951	0.952	0.952	0.953	0.953	0.954	0.954	0.955	0.955	0.956
73	0.956	0.957	0.957	0.958	0.958	0.959	0.959	0.960	0.960	0.961
74	0.961	0.962	0.962	0.963	0.963	0.964	0.964	0.965	0.965	0.965
75	0.966	0.966	0.967	0.967	0.968	0.968	0.969	0.969	0.969	0.970
76	0.970	0.971	0.971	0.972	0.972	0.972	0.973	0.973	0.974	0.974
77	0.974	0.975	0.975	0.976	0.976	0.976	0.977	0.977	0.977	0.978
78	0.978	0.979	0.979	0.979	0.980	0.980	0.980	0.981	0.981	0.981
79	0.982	0.982	0.982	0.983	0.983	0.983	0.984	0.984	0.984	0.985
80	0.985	0.985	0.985	0.986	0.986	0.986	0.987	0.987	0.987	0.987
81	0.988	0.988	0.988	0.988	0.989	0.989	0.989	0.990	0.990	0.990
82	0.990	0.991	0.991	0.991	0.991	0.991	0.992	0.992	0.992	0.992
83	0.993	0.993	0.993	0.993	0.993	0.994	0.994	0.994	0.994	0.994
84	0.995	0.995	0.995	0.995	0.995	0.995	0.996	0.996	0.996	0.996
85	0.996	0.996	0.996	0.997	0.997	0.997	0.997	0.997	0.997	0.997
86	0.998	0.998	0.998	0.998	0.998	0.998	0.998	0.998	0.998	0.999
87	0.999	0.999	0.999	0.999	0.999	0.999	0.999	0.999	0.999	0.999
88	0.999	0.999	1.000	1.000	1.000	1.000	1.000	1.000	1.000	1.000
89	1.000	1.000	1.000	1.000	1.000	1.000	1.000	1.000	1.000	1.000
90	1.000									

NATURAL COSINES

°	0′ 0.0°	6′ 0.1°	12′ 0.2°	18′ 0.3°	24′ 0.4°	30′ 0.5°	36′ 0.6°	42′ 0.7°	48′ 0.8°	54′ 0.9°
0	1.000	1.000	1.000	1.000	1.000	1.000	1.000	1.000	1.000	1.000
1	1.000	1.000	1.000	1.000	1.000	1.000	1.000	1.000	1.000	0.999
2	0.999	0.999	0.999	0.999	0.999	0.999	0.999	0.999	0.999	0.999
3	0.999	0.999	0.998	0.998	0.998	0.998	0.998	0.998	0.998	0.998
4	0.998	0.997	0.997	0.997	0.997	0.997	0.997	0.997	0.996	0.996
5	0.996	0.996	0.996	0.996	0.996	0.995	0.995	0.995	0.995	0.995
6	0.995	0.994	0.994	0.994	0.994	0.994	0.993	0.993	0.993	0.993
7	0.993	0.992	0.992	0.992	0.992	0.991	0.991	0.991	0.991	0.991
8	0.990	0.990	0.990	0.990	0.989	0.989	0.989	0.988	0.988	0.988
9	0.988	0.987	0.987	0.987	0.987	0.986	0.986	0.986	0.985	0.985
10	0.985	0.985	0.984	0.984	0.984	0.983	0.983	0.983	0.982	0.982
11	0.982	0.981	0.981	0.981	0.980	0.980	0.980	0.979	0.979	0.979
12	0.978	0.978	0.977	0.977	0.977	0.976	0.976	0.976	0.975	0.975
13	0.974	0.974	0.974	0.973	0.973	0.972	0.972	0.972	0.971	0.971
14	0.970	0.970	0.969	0.969	0.969	0.968	0.968	0.967	0.967	0.966
15	0.966	0.965	0.965	0.965	0.964	0.964	0.963	0.963	0.962	0.962
16	0.961	0.961	0.960	0.960	0.959	0.959	0.958	0.958	0.957	0.957
17	0.956	0.956	0.955	0.955	0.954	0.954	0.953	0.953	0.952	0.952
18	0.951	0.951	0.950	0.949	0.949	0.948	0.948	0.947	0.947	0.946
19	0.946	0.945	0.944	0.944	0.943	0.943	0.942	0.941	0.941	0.940
20	0.940	0.939	0.938	0.938	0.937	0.937	0.936	0.935	0.935	0.934
21	0.934	0.933	0.932	0.932	0.931	0.930	0.930	0.929	0.928	0.928
22	0.927	0.927	0.926	0.925	0.925	0.924	0.923	0.923	0.922	0.921
23	0.921	0.920	0.919	0.918	0.918	0.917	0.916	0.916	0.915	0.914
24	0.914	0.913	0.912	0.911	0.911	0.910	0.909	0.909	0.908	0.907
25	0.906	0.906	0.905	0.904	0.903	0.903	0.902	0.901	0.900	0.900
26	0.899	0.898	0.897	0.896	0.896	0.895	0.894	0.893	0.893	0.892
27	0.891	0.890	0.889	0.889	0.888	0.887	0.886	0.885	0.885	0.884
28	0.883	0.882	0.881	0.880	0.880	0.879	0.878	0.877	0.876	0.875
29	0.875	0.874	0.873	0.872	0.871	0.870	0.869	0.869	0.868	0.867
30	0.866	0.865	0.864	0.863	0.863	0.862	0.861	0.860	0.859	0.858
31	0.857	0.856	0.855	0.854	0.854	0.853	0.852	0.851	0.850	0.849
32	0.848	0.847	0.846	0.845	0.844	0.843	0.842	0.842	0.841	0.840
33	0.839	0.838	0.837	0.836	0.835	0.834	0.833	0.832	0.831	0.830
34	0.829	0.828	0.827	0.826	0.825	0.824	0.823	0.822	0.821	0.820
35	0.819	0.818	0.817	0.816	0.815	0.814	0.813	0.812	0.811	0.810
36	0.809	0.808	0.807	0.806	0.805	0.804	0.803	0.802	0.801	0.800
37	0.799	0.798	0.797	0.795	0.794	0.793	0.792	0.791	0.790	0.789
38	0.788	0.787	0.786	0.785	0.784	0.783	0.782	0.780	0.779	0.778
39	0.777	0.776	0.775	0.774	0.773	0.772	0.771	0.769	0.768	0.767
40	0.766	0.765	0.764	0.763	0.762	0.760	0.759	0.758	0.757	0.756
41	0.755	0.754	0.752	0.751	0.750	0.749	0.748	0.747	0.745	0.744
42	0.743	0.742	0.741	0.740	0.738	0.737	0.736	0.735	0.734	0.733
43	0.731	0.730	0.729	0.728	0.727	0.725	0.724	0.723	0.722	0.721
44	0.719	0.718	0.717	0.716	0.714	0.713	0.712	0.711	0.710	0.708

NATURAL COSINES

°	0' 0.0°	6' 0.1°	12' 0.2°	18' 0.3°	24' 0.4°	30' 0.5°	36' 0.6°	42' 0.7°	48' 0.8°	54' 0.9°
45	0.707	0.706	0.705	0.703	0.702	0.701	0.700	0.698	0.697	0.696
46	0.695	0.693	0.692	0.691	0.690	0.688	0.687	0.686	0.685	0.683
47	0.682	0.681	0.679	0.678	0.677	0.676	0.674	0.673	0.672	0.670
48	0.669	0.668	0.667	0.665	0.664	0.663	0.661	0.660	0.659	0.657
49	0.656	0.655	0.653	0.652	0.651	0.649	0.648	0.647	0.645	0.644
50	0.643	0.641	0.640	0.639	0.637	0.636	0.635	0.633	0.632	0.631
51	0.629	0.628	0.627	0.625	0.624	0.623	0.621	0.620	0.618	0.617
52	0.616	0.614	0.613	0.612	0.610	0.609	0.607	0.606	0.605	0.603
53	0.602	0.600	0.599	0.598	0.596	0.595	0.593	0.592	0.591	0.589
54	0.588	0.586	0.585	0.584	0.582	0.581	0.579	0.578	0.576	0.575
55	0.574	0.572	0.571	0.569	0.568	0.566	0.565	0.564	0.562	0.561
56	0.559	0.558	0.556	0.555	0.553	0.552	0.550	0.549	0.548	0.546
57	0.545	0.543	0.542	0.540	0.539	0.537	0.536	0.534	0.533	0.531
58	0.530	0.528	0.527	0.525	0.524	0.523	0.521	0.520	0.518	0.517
59	0.515	0.514	0.512	0.511	0.509	0.508	0.506	0.505	0.503	0.502
60	0.500	0.498	0.497	0.495	0.494	0.492	0.491	0.489	0.488	0.486
61	0.485	0.483	0.482	0.480	0.479	0.477	0.476	0.474	0.473	0.471
62	0.469	0.468	0.466	0.465	0.463	0.462	0.460	0.459	0.457	0.456
63	0.454	0.452	0.451	0.449	0.448	0.446	0.445	0.443	0.442	0.440
64	0.438	0.437	0.435	0.434	0.432	0.431	0.429	0.427	0.426	0.424
65	0.423	0.421	0.419	0.418	0.416	0.415	0.413	0.412	0.410	0.408
66	0.407	0.405	0.404	0.402	0.400	0.399	0.397	0.396	0.394	0.392
67	0.391	0.389	0.388	0.386	0.384	0.383	0.381	0.379	0.378	0.376
68	0.375	0.373	0.371	0.370	0.368	0.367	0.365	0.363	0.362	0.360
69	0.358	0.357	0.355	0.353	0.352	0.350	0.349	0.347	0.345	0.344
70	0.342	0.340	0.339	0.337	0.335	0.334	0.332	0.331	0.329	0.327
71	0.326	0.324	0.322	0.321	0.319	0.317	0.316	0.314	0.312	0.311
72	0.309	0.307	0.306	0.304	0.302	0.301	0.299	0.297	0.296	0.294
73	0.292	0.291	0.289	0.287	0.286	0.284	0.282	0.281	0.279	0.277
74	0.276	0.274	0.272	0.271	0.269	0.267	0.266	0.264	0.262	0.261
75	0.259	0.257	0.255	0.254	0.252	0.250	0.249	0.247	0.245	0.244
76	0.242	0.240	0.239	0.237	0.235	0.233	0.232	0.230	0.228	0.227
77	0.225	0.223	0.222	0.220	0.218	0.216	0.215	0.213	0.211	0.210
78	0.208	0.206	0.205	0.203	0.201	0.199	0.198	0.196	0.194	0.193
79	0.191	0.189	0.187	0.186	0.184	0.182	0.181	0.179	0.177	0.175
80	0.174	0.172	0.170	0.168	0.167	0.165	0.163	0.162	0.160	0.158
81	0.156	0.155	0.153	0.151	0.150	0.148	0.146	0.144	0.143	0.141
82	0.139	0.137	0.136	0.134	0.132	0.131	0.129	0.127	0.125	0.124
83	0.122	0.120	0.118	0.117	0.115	0.113	0.111	0.110	0.108	0.106
84	0.105	0.103	0.101	0.099	0.098	0.096	0.094	0.092	0.091	0.089
85	0.087	0.085	0.084	0.082	0.080	0.078	0.077	0.075	0.073	0.072
86	0.070	0.068	0.066	0.065	0.063	0.061	0.059	0.058	0.056	0.054
87	0.052	0.051	0.049	0.047	0.045	0.044	0.042	0.040	0.038	0.037
88	0.035	0.033	0.031	0.030	0.028	0.026	0.024	0.023	0.021	0.019
89	0.017	0.016	0.014	0.012	0.010	0.009	0.007	0.005	0.003	0.002
90	0.000									

NATURAL TANGENTS

°	0' 0.0°	6' 0.1°	12' 0.2°	18' 0.3°	24' 0.4°	30' 0.5°	36' 0.6°	42' 0.7°	48' 0.8°	54' 0.9°
0	0.000	0.002	0.003	0.005	0.007	0.009	0.010	0.012	0.014	0.016
1	0.017	0.019	0.021	0.023	0.024	0.026	0.028	0.030	0.031	0.033
2	0.035	0.037	0.038	0.040	0.042	0.044	0.045	0.047	0.049	0.051
3	0.052	0.054	0.056	0.058	0.059	0.061	0.063	0.065	0.066	0.068
4	0.070	0.072	0.073	0.075	0.077	0.079	0.080	0.082	0.084	0.086
5	0.087	0.089	0.091	0.093	0.095	0.096	0.098	0.100	0.102	0.103
6	0.105	0.107	0.109	0.110	0.112	0.114	0.116	0.117	0.119	0.121
7	0.123	0.125	0.126	0.128	0.130	0.132	0.133	0.135	0.137	0.139
8	0.141	0.142	0.144	0.146	0.148	0.149	0.151	0.153	0.155	0.157
9	0.158	0.160	0.162	0.164	0.166	0.167	0.169	0.171	0.173	0.175
10	0.176	0.178	0.180	0.182	0.184	0.185	0.187	0.189	0.191	0.193
11	0.194	0.196	0.198	0.200	0.202	0.203	0.205	0.207	0.209	0.211
12	0.213	0.214	0.216	0.218	0.220	0.222	0.224	0.225	0.227	0.229
13	0.231	0.233	0.235	0.236	0.238	0.240	0.242	0.244	0.246	0.247
14	0.249	0.251	0.253	0.255	0.257	0.259	0.260	0.262	0.264	0.266
15	0.268	0.270	0.272	0.274	0.275	0.277	0.279	0.281	0.283	0.285
16	0.287	0.289	0.291	0.292	0.294	0.296	0.298	0.300	0.302	0.304
17	0.306	0.308	0.310	0.311	0.313	0.315	0.317	0.319	0.321	0.323
18	0.325	0.327	0.329	0.331	0.333	0.335	0.337	0.338	0.340	0.342
19	0.344	0.346	0.348	0.350	0.352	0.354	0.356	0.358	0.360	0.362
20	0.364	0.366	0.368	0.370	0.372	0.374	0.376	0.378	0.380	0.382
21	0.384	0.386	0.388	0.390	0.392	0.394	0.396	0.398	0.400	0.402
22	0.404	0.406	0.408	0.410	0.412	0.414	0.416	0.418	0.420	0.422
23	0.424	0.427	0.429	0.431	0.433	0.435	0.437	0.439	0.441	0.443
24	0.445	0.447	0.449	0.452	0.454	0.456	0.458	0.460	0.462	0.464
25	0.466	0.468	0.471	0.473	0.475	0.477	0.479	0.481	0.483	0.486
26	0.488	0.490	0.492	0.494	0.496	0.499	0.501	0.503	0.505	0.507
27	0.510	0.512	0.514	0.516	0.518	0.521	0.523	0.525	0.527	0.529
28	0.532	0.534	0.536	0.538	0.541	0.543	0.545	0.547	0.550	0.552
29	0.554	0.557	0.559	0.561	0.563	0.566	0.568	0.570	0.573	0.575
30	0.577	0.580	0.582	0.584	0.587	0.589	0.591	0.594	0.596	0.598
31	0.601	0.603	0.606	0.608	0.610	0.613	0.615	0.618	0.620	0.622
32	0.625	0.627	0.630	0.632	0.635	0.637	0.640	0.642	0.644	0.647
33	0.649	0.652	0.654	0.657	0.659	0.662	0.664	0.667	0.669	0.672
34	0.675	0.677	0.680	0.682	0.685	0.687	0.690	0.692	0.695	0.698
35	0.700	0.703	0.705	0.708	0.711	0.713	0.716	0.719	0.721	0.724
36	0.727	0.729	0.732	0.735	0.737	0.740	0.743	0.745	0.748	0.751
37	0.754	0.756	0.759	0.762	0.765	0.767	0.770	0.773	0.776	0.778
38	0.781	0.784	0.787	0.790	0.793	0.795	0.798	0.801	0.804	0.807
39	0.810	0.813	0.816	0.818	0.821	0.824	0.827	0.830	0.833	0.836
40	0.839	0.842	0.845	0.848	0.851	0.854	0.857	0.860	0.863	0.866
41	0.869	0.872	0.875	0.879	0.882	0.885	0.888	0.891	0.894	0.897
42	0.900	0.904	0.907	0.910	0.913	0.916	0.920	0.923	0.926	0.929
43	0.933	0.936	0.939	0.942	0.946	0.949	0.952	0.956	0.959	0.962
44	0.996	0.969	0.972	0.976	0.979	0.983	0.986	0.990	0.993	0.997

NATURAL TANGENTS

°	0' 0.0°	6' 0.1°	12' 0.2°	18' 0.3°	24' 0.4°	30' 0.5°	36' 0.6°	42' 0.7°	48' 0.8°	54' 0.9°
45	1.000	1.003	1.007	1.011	1.014	1.018	1.021	1.025	1.028	1.032
46	1.036	1.039	1.043	1.046	1.050	1.054	1.057	1.061	1.065	1.069
47	1.072	1.076	1.080	1.084	1.087	1.091	1.095	1.099	1.103	1.107
48	1.111	1.115	1.118	1.122	1.126	1.130	1.134	1.138	1.142	1.146
49	1.150	1.154	1.159	1.163	1.167	1.171	1.175	1.179	1.183	1.188
50	1.192	1.196	1.200	1.205	1.209	1.213	1.217	1.222	1.226	1.230
51	1.235	1.239	1.244	1.248	1.253	1.257	1.262	1.266	1.271	1.275
52	1.280	1.285	1.289	1.294	1.299	1.303	1.308	1.313	1.317	1.322
53	1.327	1.332	1.337	1.342	1.346	1.351	1.356	1.361	1.366	1.371
54	1.376	1.381	1.387	1.392	1.397	1.402	1.407	1.412	1.418	1.423
55	1.428	1.433	1.439	1.444	1.450	1.455	1.460	1.466	1.471	1.477
56	1.483	1.488	1.494	1.499	1.505	1.511	1.517	1.522	1.528	1.534
57	1.540	1.546	1.552	1.558	1.564	1.570	1.576	1.582	1.588	1.594
58	1.600	1.607	1.613	1.619	1.625	1.632	1.638	1.645	1.651	1.658
59	1.664	1.671	1.677	1.684	1.691	1.698	1.704	1.711	1.718	1.725
60	1.732	1.739	1.746	1.753	1.760	1.767	1.775	1.782	1.789	1.797
61	1.804	1.811	1.819	1.827	1.834	1.842	1.849	1.857	1.865	1.873
62	1.881	1.889	1.897	1.905	1.913	1.921	1.929	1.937	1.946	1.954
63	1.963	1.971	1.980	1.988	1.997	2.006	2.014	2.023	2.032	2.041
64	2.050	2.059	2.069	2.078	2.087	2.097	2.106	2.115	2.125	2.135
65	2.144	2.154	2.164	2.174	2.184	2.194	2.204	2.215	2.225	2.236
66	2.246	2.257	2.267	2.278	2.289	2.300	2.311	2.322	2.333	2.344
67	2.356	2.367	2.379	2.391	2.402	2.414	2.426	2.438	2.450	2.463
68	2.475	2.488	2.500	2.513	2.526	2.539	2.552	2.565	2.578	2.592
69	2.605	2.619	2.632	2.646	2.660	2.675	2.689	2.703	2.718	2.733
70	2.747	2.762	2.778	2.793	2.808	2.824	2.840	2.855	2.872	2.888
71	2.904	2.921	2.937	2.954	2.971	2.989	3.006	3.024	3.041	3.059
72	3.078	3.096	3.115	3.133	3.152	3.172	3.191	3.211	3.230	3.250
73	3.271	3.291	3.312	3.333	3.354	3.376	3.398	3.420	3.442	3.464
74	3.487	3.510	3.534	3.557	3.582	3.606	3.630	3.655	3.681	3.706
75	3.732	3.758	3.785	3.812	3.839	3.867	3.895	3.923	3.952	3.981
76	4.011	4.041	4.071	4.102	4.133	4.165	4.198	4.230	4.263	4.297
77	4.331	4.366	4.401	4.437	4.474	4.511	4.548	4.586	4.625	4.665
78	4.704	4.745	4.787	4.829	4.871	4.915	4.959	5.004	5.050	5.097
79	5.145	5.193	5.242	5.292	5.343	5.395	5.449	5.502	5.558	5.614
80	5.671	5.729	5.789	5.850	5.912	5.976	6.040	6.107	6.174	6.243
81	6.313	6.386	6.459	6.535	6.612	6.691	6.772	6.855	6.939	7.026
82	7.115	7.206	7.300	7.396	7.495	7.595	7.699	7.806	7.916	8.028
83	8.144	8.263	8.386	8.512	8.642	8.776	8.915	9.058	9.205	9.357
84	9.514	9.677	9.844	10.02	10.20	10.39	10.58	10.78	10.99	11.20
85	11.43	11.66	11.91	12.16	12.43	12.71	13.00	13.30	13.62	13.95
86	14.30	14.67	15.06	15.46	15.89	16.35	16.83	17.34	17.88	18.46
87	19.08	19.74	20.45	21.20	22.02	22.90	23.86	24.90	26.03	27.27
88	28.64	30.14	31.82	33.69	35.80	38.19	40.92	44.07	47.74	52.08
89	57.29	63.66	71.62	81.85	95.49	114.6	143.2	191.0	286.5	573.0
90										